TOPICS AT THE FRONTIER OF STATISTICS AND NETWORK ANALYSIS

(Re)Visiting the Foundations

Eric D. Kolaczyk

SEMSTAT ELEMENTS

Ernst C. Wit

The Bernoulli Society

CAMBRIDGE
UNIVERSITY PRESS

Cambridge Elements

CAMBRIDGE
UNIVERSITY PRESS

University Printing House, Cambridge CB2 8BS, United Kingdom

One Liberty Plaza, 20th Floor, New York, NY 10006, USA

477 Williamstown Road, Port Melbourne, VIC 3207, Australia

4843/24, 2nd Floor, Ansari Road, Daryaganj, Delhi – 110002, India

79 Anson Road, #06–04/06, Singapore 079906

Cambridge University Press is part of the University of Cambridge.

It furthers the University's mission by disseminating knowledge in the pursuit of education, learning, and research at the highest international levels of excellence.

www.cambridge.org
Information on this title: www.cambridge.org/9781108407120
DOI: 10.1017/9781108290159

First published 2017

A catalogue record for this publication is available from the British Library.

ISBN 978-1-108-40712-0 Paperback
ISSN 2514-3778 (Print)
ISSN 2398-404X (Online)

"I will not follow where the path may lead, but I will go where there is no path, and I will leave a trail."

Muriel Strode, from *Wind-Wafted Wild Flowers*

To the many people in my life who have, in one way or another, chosen to push 'the frontier,' leaving a trail rather than following the common path. Thank you – for your courage and your example. – EDK

Contents

Illustrations

Tables

Preface

It is an honor and a pleasure to be asked to contribute this monograph, and the associated lectures, to the Séminaire Européen de Statistique (SemStat) series on new developments in statistics. While network analysis itself is not new – having exploded onto the scientific scene at least 15 years ago – its interface with statistics arguably now represents one of the most active and exciting frontiers of our field. In a little over a decade, we have gone from there being only a handful of statistics researchers deeply involved with network-based research to a point where articles on network-related topics appear regularly in the top statistics journals.

Ironically, however, much of this work is foundational in nature. As of this writing, Google Scholar lists approximately 500,000 articles published in the sciences and humanities since 1999 with the word *network* in the title. And a significant proportion of that work involves statistical analysis of network data of one type or another. Yet, while some of that work certainly addresses foundational topics, much of the statistical foundations of network analysis remain to be laid. In fact, they are being laid right now.

By 'foundational' I mean topics like modeling, sampling, and design – the three main topics featured in this monograph. These are topics whose development is continually revisited by researchers in our field as we encounter fundamentally new data structures. Hence the subtitle. My goal is to provide a snapshot of the current frontier of statistics and network analysis, in the context of such topics. Emphasis is not only on what has been done, but on what remains to be done. As such this monograph will likely be of interest to, and indeed is written primarily for, graduate students and researchers in statistics and closely related fields.

I would like to thank Ernst Wit and the SemStat committee for the invitation to contribute to this series. I would also like to thank Edo Airoldi, Harry Crane, Juliane Manitz, and Dan Sussman for kindly giving feedback on various aspects of this monograph. Finally, I would like to acknowledge the support of my own research work on topics herein through grants from the U.S. Air Force Office of Scientific Research (AFOSR) and the U.S. Army Research Office (ARO).

1
Background and Overview

Network analysis (i.e., the analysis of network-indexed systems) was a relatively small field of study – if it can even be said to have been a field previously – until about 15 years ago. Around the turn of the century, there was an explosion of interest in networks and their analysis, which has continued to propagate ever since. Early on during this period, questions and problems in fields like computer network traffic analysis, computational biology, and social media drove much of the interest in networks.[1] That interest has continued to spread in more recent years to various other fields like finance and computational neuroscience.

It is tempting to argue that this fascination with networks reflects increasingly a shift towards more of a systems-level perspective in science, perhaps in contrast to the more reductionist perspective of much of the previous century. But such a question is perhaps best left to those studying the history of science. Critically important from the statistical point of view, however, has been the corresponding movement towards simultaneous measurement of the various components in the systems around us. The tendency towards so-called high-throughput measurement in biology is an example. It is the availability of data from such measurements that both gives rise to the need for and suggests the potential usefulness of statistics *per se* in network analysis.

In referring to the statistical analysis of network data, we have in mind analysis of data either *of* or *from* a system conceptualized as a network. There is a dizzying array of examples of such systems, data, and analyses to be found now in the literature. In the context of engineered networks, traffic volume measurements are monitored for anomalies in communication networks; utilization data in energy

[1] Some of which was motivated and inspired in turn by interest going even further back in parts of social science research, in the form of social network analysis.

networks are used to predict demand and set pricing; and environmental readings in sensor networks drive policy in managing natural resources. Within biology we seek to understand the connectivity of structural and functional brain networks, in relation to everything from disease to everyday activities; experimentally verified affinity for binding between proteins is used to predict both additional protein–protein interactions and protein function; and information on predator–prey relationships allows for detailed simulation of population dynamics within a given ecosystem. Finally, within the world of social media there is keen interest in detecting communities in online social networks like Facebook; in understanding how news and opinion propagates through platforms like Twitter; and in optimally perturbing product–purchaser relations in ecommerce settings, through the construction of recommender systems like those pioneered by Amazon.

In analyzing network data we can find ourselves confronted with analogues of many (if not all) of the same fundamental types of problems encountered in a standard 'Statistics 101' course taught to college freshmen. These include problems of sampling and design, description, modeling and inference, and prediction, among others. The Oxford English Dictionary defines a *network* to be simply 'a collection of interconnected things.' It is the relational aspect of network data that lends to network analysis an element of uniqueness. This relational aspect typically manifests as a type of complex dependency which, importantly, cannot be mapped directly and exclusively to, for example, the more familiar notions of temporal or spatial dependencies, although certainly the latter often play a role. Simply put, there is in network analysis both the familiar and the new.

There is much that has been done in 15 years of work on network analysis. And a substantial amount of this work has been statistical in nature. So how then can we claim that this monograph is about the 'frontiers'? It is in large part because statistics as a field, as opposed to a number of other similarly quantitative fields, has only recently begun to engage with network analysis deeply and *en masse*. This delay, in turn, is due at least partly to the fact that it has taken some time for network analysis itself to be perceived as a field, as opposed to just a fad. Regardless of the reasons, much of the statistical foundations in network analysis have yet to be laid. But this state of affairs is rapidly changing. In recent years, the pervasiveness of network-based research throughout the sciences has increasingly demanded the attention of our field. At the same time, the combination of new and familiar mentioned above has attracted that same attention. Further accelerating this shift in focus is the fact that many network analysis problems also touch on other current themes of interest in our field, such as high-dimensionality (particularly 'small n, large p' problems), sparseness, analysis of unstructured data, and 'big data.'

Our goal in this monograph is to present a brief introduction to foundational topics currently at the frontier of statistics and network analysis. The treatment here is of a summary form, and primarily technical in nature. For more of an overview of the statistical analysis of network data in general, the reader is referred to [127, 128]. The latter reference includes an introduction to various packages in R for doing network analysis.

There are four chapters following this one, covering three primary topic areas in the first three, and then closing with a compilation of emerging topic areas in the last. We begin in Chapter 2 with the modeling of an observed network. We focus specifically on a class of so-called latent network models, wherein arguably the majority of recent work in statistics on network modeling has taken place. There we assess the progress made in this area to date against the standard of the classical linear modeling framework, observing that while much has been done, much remains still to be done.

We then turn our attention, in Chapter 3, to sampling from networks and the corresponding problem of inferring network characteristics from a sample. Curiously, in this area a good deal of seminal work was done in the 1970s and 1980s, after which little activity was apparent until the modern era. While much energy since has been devoted to devising sampling plans that allow standard estimators to perform well for given network characteristic(s), much less energy has been spent on the problem of devising better estimators adapted to a given sampling plan. We will look specifically at progress and challenges in sampling and estimation of characteristics having to do with vertex degree.

The similarly classical area of experimental design is the topic of Chapter 4. There we describe the recently emerging area of networked experiments (i.e., experiments in populations of units whose response to treatment or intervention bears an interdependence that can be represented using a network). Both the design and analysis of such networked experiments is of intense interest in the social and economic sciences, and increasingly, as well, in the biomedical sciences. Interestingly, progress in this area seems likely to also yield progress in the related field of causal inference, where recent work suggests that a network-based perspective is especially useful for capturing and dealing with the notion of interference in the classical potential outcomes framework.

Finally, in Chapter 5, we visit briefly a number of more nascent topics, each of which nevertheless appears likely to provide fertile ground for substantial growth and development in statistics research. These consist of the notion of dynamic and multi-networks (i.e., networks represented by graphs evolving in time or, more generally, consisting of multiple layers); the issue of noisy networks (i.e., where there is uncertainty in the status of at least some of the (non)edges); and the setting in which we observe many networks (i.e., in contrast to the rest of this monograph, where the perspective is that of analyzing a single network).

We have also included a brief appendix. When speaking with statisticians about network analysis over the years, we have found that, naturally, it is not a lack of statistical knowledge that is the barrier to deeper involvement, but rather it is typically a lack of knowledge of networks. In particular, a lack of knowledge of basics about the theory, representation, and computation of graphs. Accordingly, the appendix contains some necessary (and, hopefully, minimally sufficient!) background in this regard. This treatment should be enough to allow graduate students and researchers trained in statistics, or closely related fields, to quickly and efficiently access the material herein.

In closing, it is worth noting that this monograph certainly does not by any means attempt to survey the entire frontier of statistics and network analysis. There are many topics that are currently of active interest in either the statistics or greater network science literatures – or both – which are not addressed here. This includes, perhaps foremost, the problem of inferring a network topology (a.k.a., 'network inference'), which arguably is already a fairly mature topic area, due to its close connections with high-dimensional regression. But, as Gary Keller has written in his book *The One Thing: The Surprisingly Simple Truth Behind Extraordinary Results*, "It is not that we have too little time to do all the things we need to do, it is that we feel the need to do too many things in the time we have." While we make no claims as to the extraordinariness of this modest monograph, we have nevertheless tried to heed the wisdom of these words in its writing!

2
Network Modeling

2.1 Introduction

Suppose we observe a network. That is, we observe a collection of elements in a system of interest and the presence/absence of some notion of a relationship between pairs of elements. It is typical to associate that network with a graph $G = (V, E)$, where $V = \{1, \ldots, N_v\}$ is a set of vertices and E is a set of edges. Throughout this monograph, unless stated otherwise, we will assume G to be undirected, unweighted, and simple (i.e., without multi-edges or loops). Corresponding to this graph is the $N_v \times N_v$ adjacency matrix, which we will denote by $\mathbf{Y}$, where $Y_{ij} = 1$ if and only if $\{i,j\} \in E$.

One of the tasks of most fundamental interest in statistical network analysis has been the modeling of an observed network graph. By a model, we mean effectively a collection $\{\mathbb{P}_\theta(G), G \in \mathcal{G} : \theta \in \Theta\}$, where $\mathcal{G}$ is a set (or 'ensemble') of possible graphs, $\mathbb{P}_\theta$ is a probability distribution on $\mathcal{G}$, and θ is a vector of parameters, ranging over possible values in Θ. Equivalently, we can think of a model $\mathbb{P}_\theta(\mathbf{Y})$ over adjacency matrices $\mathbf{Y}$. In practice, such models are used for various purposes, including (i) testing for 'significance' of pre-defined characteristic(s) in a given network graph, (ii) the study of proposed mechanisms for generating certain properties felt to be canonical (or at least common) in real-world networks (such as broad degree distributions or small-world effects), and (iii) the assessment of factors potentially predictive of relational ties.

Contributions to network modeling have come from many fields. These include mathematics and probability theory (e.g., [32]), statistical physics (e.g., [154, Sec IV]), and social network analysis (e.g., [143]). Each traditionally has had its own particular emphasis and goals. Modern statistical treatment of network modeling has tended to draw on aspects of all three of these fields. Fienberg [74] provides a recent history of statistical models for network analysis.

Two key approaches to network modeling in statistics have been (a) through conditional independencies, via a graphical models perspective, and (b) through the use of latent variables. The primary example of the former is the class of exponential random graph models (ERGMs), and of the latter, the class of stochastic block models (SBMs). We will focus on the latter, for which arguably the most complete development exists, due in no small part to the fact that SBMs are, compared to ERGMs, simpler in structure and more amenable to theoretical analysis. In addition, ERGMs, while more heavily developed earlier on, have in recent years come upon certain stumbling blocks involving model degeneracy that are still the subject of ongoing study (e.g., [46, 100]).

So why is statistical network modeling considered a 'frontier' topic for the purposes of this monograph? In a 2007 article, Robins and Morris [163], offered the following as a goal in this area:

> *A good [statistical network] model needs to be both estimable from data and a reasonable representation of that data, to be theoretically plausible about the type of effects that might have produced the network, and to be amenable to examining which competing effects might be the best explanation of the data.*

This statement is reflective of the type of standards already familiar to most graduate students after, say, a thorough course in classical linear modeling – standards which have been met in numerous other statistical modeling contexts. We are still, however, in the early stages of meeting these standards with network modeling. Just as other key dependent-data paradigms, such as time series analysis and spatial data analysis, required ample time to mature to a state of knowledge comparable to more established paradigms for independent and identically distributed data, the relational nature of network data is sufficiently different from these other paradigms that it has proven to need its own maturation period as well.

At the time of this writing, work on latent network models has in just the last few years allowed us to come close to achieving the goals stated above – at least when viewed as an initial pass or a proof-of-concept. The goal of this chapter is to introduce two related classes of latent network models – the parametric stochastic block model and a nonparametric analogue – and to detail the extent to which these models meet the demands we place upon them, outlining both progress and current challenges in the area.

2.2 Latent Network Models: A Canonical Class

The use of latent variables in modeling observed data is a common and often-powerful technique in statistics generally. Latent network models incorporate the same principle in the context of statistical network analysis. Two popular choices for the type of latent variables used in network modeling are latent classes and latent features. In the first case, latent class membership is assumed to drive the propensity towards establishing relational ties. In the second case, relational ties are viewed as more likely to be formed between those that are somehow 'sufficiently close' in some space of (typically continuous) vertex characteristics. Some or all of this characteristic space is assumed unobserved, and hence latent.

There are by now many variations on the notion of latent network models. Here we focus on just two: a class of parametric models known as stochastic block models, and a more general and nonparametric analogue, based on the notion of a so-called graphon. The connection between the two has, in particular, proven to yield a powerful lever for theoretical statistical analysis.

2.2.1 *Stochastic Block Models*

Suppose that each vertex $i \in V$ of a graph $G = (V, E)$ can belong to one of Q classes, say $\mathcal{C}_1, \ldots, \mathcal{C}_Q$. And furthermore, suppose that we know the class label $q = q(i)$ for each vertex i. A *block model* for G specifies that each element Y_{ij} of the adjacency matrix $\mathbf{Y}$ is, conditional on the class labels q and r of vertices i and j, respectively, an independent Bernoulli random variable, with probability π_{qr}. For an undirected graph, $\pi_{qr} = \pi_{rq}$. The block model is hence a variant of the classical Bernoulli random graph model,[1] where the probability of an edge can be one of Q^2 possible values π_{qr}.

The assumption that the class membership of vertices is known or, moreover, that the 'true' classes $\mathcal{C}_1, \ldots, \mathcal{C}_Q$ have been correctly specified, is generally considered untenable in practice. More common, therefore, is the use of a *stochastic block model* (SBM). Pioneered by Holland and colleagues [109], this model specifies only that there are

[1] A classical random graph model assigns edges independently between N_v vertices with common probability π. The study of such models was initiated through a series of seminal papers by Erdős and Rényi [71, 72, 73]. See the book by Bollobás [32], for example, for a comprehensive summary of known results for such models.

Q classes, for some Q, but does not specify the nature of those classes nor the class membership of the individual vertices. Rather, it dictates simply that the latent class membership of each vertex i be determined independently, according to a common distribution on the set $\{1,\ldots,Q\}$.

Formally, let $Z_{iq} = 1$ if vertex i is of class q, and zero otherwise. Under a stochastic block model, each vector $\mathbf{Z}_i = (Z_{i1},\ldots,Z_{iQ})$ has one and only one nonzero element, the determination of which is made independently for each i according to probabilities $\mathbb{P}(Z_{iq} = 1) = \alpha_q$, where $\sum_{q=1}^{Q} \alpha_q = 1$. Then, conditional on the values $\{\mathbf{Z}_i\}$, the entries Y_{ij} are again modeled as independent Bernoulli random variables, with probabilities π_{qr}, as in the nonstochastic block model. Expressed concisely, the stochastic block model can be written as

$$\mathbf{Z}_i \overset{i.i.d.}{\sim} \text{Multinomial}(1,\alpha), \tag{2.1}$$

$$Y_{ij}|\mathbf{Z}_i = \mathbf{z}_i, \mathbf{Z}_j = \mathbf{z}_j \sim \text{Bernoulli}(\pi_{\mathbf{z}_i,\mathbf{z}_j}), \tag{2.2}$$

for $1 \le i,j \le N_v$, where $Y_{ij} = Y_{ji}$, $Y_{ii} \equiv 0$, and $\pi_{\mathbf{z}_i,\mathbf{z}_j}$ is to be understood as π_{qr} for the particular pair q, r for which $z_{iq} = 1$ and $z_{jr} = 1$.

A stochastic block model is thus, effectively, a mixture of classical random graph models. Many of the properties of such models are therefore relatively straightforward to derive. For example, the probability that there is an edge between i and j is simply

$$\mathbb{P}(Y_{ij} = 1) = \sum_{1 \le q,r \le Q} \alpha_q \alpha_r \pi_{qr}.$$

Similarly, the expectation of the average vertex degree under this model is $2M/N_v$, where $M = \sum_{i<j} \mathbb{P}(Y_{ij} = 1)$. Several other examples of similarly straightforward results may be found in [59]. A treatment of various more sophisticated characteristics of stochastic block models, such as the emergence of a giant component and the size distribution of classes C_q, may be found in a series of papers by Soderberg [181, 182, 183]. These results, in turn, have been shown to be special cases under a more general formalism for inhomogeneous random graphs developed in detail by Bollobás and colleagues [34].

Note that, given the definition of these models through mixtures, the user of stochastic block models typically will face problems of identifiability (from the perspective of statistical inference).

In general, these models are said to be identifiable only up to permutation of class labels. While essentially accurate, in fact the question of identifiability for stochastic block models is somewhat more subtle than this, requiring conditions as well on both the distinctness of the parameters α and π (not surprisingly) and, rather less evident, on the number of vertices N_v. See the work of Allman, Matias, and Rhodes [10, 11] for details.

2.2.2 Exchangeable Models Based on Graphons

The stochastic block model has the property of (vertex) exchangeability, in the sense that the distribution of $\mathbf{Y}$ under this model remains unchanged when an arbitrary permutation is applied to its rows and columns. Note that these are parametric models in the sense that they are determined up to the Q mixing weights α_q and the Q^2 connection probabilities π_{qr}.

A nonparametric variant of the model in (2.1)–(2.2) may be defined as follows:

$$U_1, \ldots, U_{N_v} \overset{i.i.d.}{\sim} \text{Uniform } (0, 1), \tag{2.3}$$

$$Y_{ij} | U_i = u_i, U_j = u_j \sim \text{Bernoulli}(f(u_i, u_j)), \tag{2.4}$$

where f: $[0, 1]^2 \rightarrow [0, 1]$ is a symmetric, measurable function called a graphon – short for 'graph function.' Also, here again, we dictate that $Y_{ij} = Y_{ji}$ and $Y_{ii} \equiv 0$. Under this model, the latent variables $\mathbf{Z}_i$ in the SBM, randomly assigning membership of vertices to one of a finite number of Q classes, have been replaced by independent and identically distributed uniform random variables, randomly assigning locations on the interval [0, 1]. The graphon, evaluated at the appropriate arguments $U_i = u_i$ and $U_j = u_j$, then plays the role played previously by the probabilities π_{qr} in the randomized decision for whether or not there is an edge between the pair of vertices i and j.

Note that we can recover the parametric SBM in (2.1)–(2.2) from the nonparametric formulation in (2.3)–(2.4) by (i) partitioning the unit interval into Q intervals, of lengths $\alpha_1, \ldots, \alpha_Q$; (ii) taking the Cartesian product of that partition with itself to partition, in turn, the unit square $[0, 1]^2$ into Q^2 blocks; and (iii) defining f to be piecewise constant on blocks, with height π_{qr} on the qrth block.

The relationship between the models in (2.1)–(2.2) and (2.3)–(2.4) is analogous to that between classical linear regression and nonparametric regression. Recall that in classical linear regression, for example, our interest is in models of the form $\mathbf{y} = \mathbf{X}\beta + \in$, where $\mathbf{X}$ is a known $n \times p$ matrix and β is an unknown vector of fixed and finite dimension p. On the other hand, in nonparametric regression, the standard model specifies that $\mathbf{y} = \mathrm{f} + \in$, where $\mathrm{f} = (f(0), f(1/n), \ldots, f((n-1)/n))^T$ is the evaluation of some unknown function f at n equispaced points on [0, 1]. (For both models, $\in$ is an additive error term.) Importantly, in the case of the latter model, the parameter of interest to us (i.e., the function f) is an infinite-dimensional object. The nonparametric random graph model in (2.3)–(2.4) is therefore quite similar in spirit to traditional nonparametric regression, but with the key distinction that the design points at which the function f is evaluated are random and unobserved.

Like the SBM, the graphon-based extension in (2.3)–(2.4) can be seen to be exchangeable, by construction. In fact, it is only a specific case of a more general model with the property of exchangeability for (not necessarily binary) random arrays. The seminal work on this topic was done independently by Aldous [9] and by Hoover [110], with further contributions by Kallenberg [122]. The celebrated Aldous–Hoover theorem pertains to two-dimensional random arrays $\mathbf{A} = [A_{ij}]_{i,j \geq 1}$, with values $A_{ij} \in \mathcal{A}$, for some sample space $\mathcal{A}$. Specifically, the theorem states that such an array $\mathbf{A}$ is exchangeable[2] if and only if there is a random function[3] $F: [0, 1]^3 \to \mathcal{A}$ such that A_{ij} is equal in distribution to $F(U_i, U_j, U_{\{ij\}})$, where the U_i and U_j are random sequences and the $U_{\{ij\}}$ a random array of independent and identically distributed uniform random variables on [0, 1]. Here, F may be thought of as a generalization of a (random) graphon. This theorem is an extension of de Finetti's classical theorem for exchangeable binary sequences. See [158] for a broad discussion of such extensions to graphs, arrays, and other related random structures.

[2] We say that a random two-dimensional array $\mathbf{A} = [A_{ij}]_{i,j \geq 1}$ is exchangeable if $\mathbf{A}$ is equal in distribution to $[A_{\sigma(i)\sigma(j)}]_{i,j \geq 1}$ for every permutation σ of the natural numbers $\mathbb{N}$.

[3] That is, under this particular variant of the Aldous–Hoover theorem, the function F itself is random. Whether or not this underlying latent structure is assumed random or fixed is an important distinction to note generally within the literature on latent network modeling, as the choice has obvious implications for questions of inference and the like.

The Aldous–Hoover theorem was first exploited for statistical modeling of networks by Hoff [107] and by Bickel and Chen [24], each of whom make various restrictions to this general formulation suitable for their purposes, arriving ultimately at models in the spirit of that defined above, in (2.3)–(2.4). Such restrictions are necessary (and not just convenient) because the parameterization of these models is not unique (i.e., the same random graph model can be parameterized by more than one distinct graphon. For example, the graphons $f(x,y)$ and $f(1-x,1-y)$, for any given choice of f, yield the same model, since U and $1-U$ are equal in distribution when $U \sim \mathrm{Uniform}(0,1)$. More generally, the same will hold true under transformation of the arguments of f by any measure-preserving transformation (i.e., any mapping $\phi : [0,1] \rightarrow [0,1]$ for which $\phi(U)$ remains uniform on $[0,1]$).

Similar to the case of stochastic block models, the implication of the above observations from a statistical perspective is that these models are not identifiable. Bickel and Chen [24] offer further restrictions, through the use of monotonization, which force identifiability on the model. The so-called strict monotonicity criterion assumes that there exists a measure-preserving map ϕ such that the induced graphon $\widetilde{f}(x,y) = f(\phi(x), \phi(y))$ has the property that the degree function[4]

$$\widetilde{g}(x) = \int_0^1 \widetilde{f}(x,y)dy \tag{2.5}$$

is strictly increasing in x. In this case, $\widetilde{f}$ is called the canonical graphon of f. Restriction to this latter graphon yields identifiability of the model. But see [158, Rmk 3.8] for cautionary remarks on this approach as a general panacea. For more detailed discussions of identifiability and additional solutions, see [62, Thm 7.1], [26, Sec 3], and [208, Sec 2]. Alternatively, the model can be considered identifiable up to equivalence classes, as in Kallenberg [121].

Figure 2.1 shows an illustration of a graphon-based random graph model for the relatively simple case of $f(x,y) = \min(x,y)$. It is straightforward to see in the visualization of f how the chance of an edge is determined. But if we are given only the adjacency matrix for the overall resulting random graph, it is impossible to recover the information on

[4] The degree function can be viewed as a generalization of the degree sequence, which in the case of a finite graph is obtained by taking row (or column) sums of the adjacency matrix.

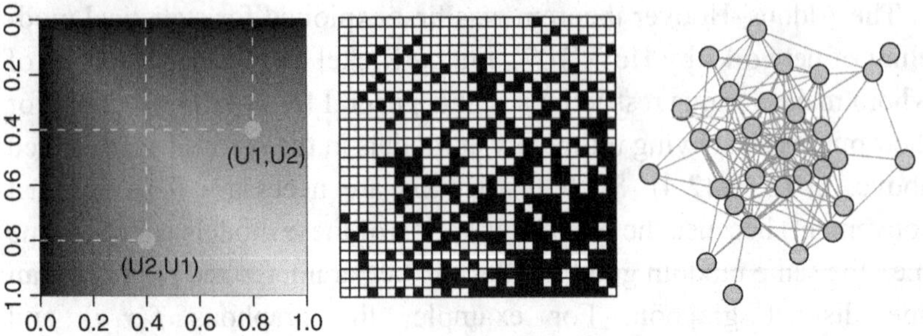

Figure 2.1 Illustration of the graphon-based, nonparametric random graph model defined in (2.3)–(2.4). *Left:* Gray-scale rendering of the graphon $f(x,y) = \min(x,y)$. Shown are a point (and its reflection) representing the realization $U_1 = 0.8$ and $U_2 = 0.4$ of two uniform random variables. An edge arises between the two corresponding vertices i and j with probability $\min(0.8, 0.4) = 0.4$. *Center:* Visualization of the adjacency matrix for a network of $N_v = 30$ vertices simulated according to the model. *Right:* Visualization of the corresponding network. (Example based on Lovász [138].)

how it was generated. Here the visualizations of the adjacency matrix and the graphon appear similar – but this is only because we arbitrarily ordered the vertices $\{1,\ldots,N_v\}$ according to the order statistics of the underlying uniform random variables, $U_{(1)},\ldots,U_{(N_v)}$.

On a final note, we point out that graphons are objects central to the literature on graph limits, a highly active area of research that has emerged just over the past decade. Indeed, it is in the context of graph limits that the term 'graphon' was coined, and the concept developed. See the book by Lovász [138] for a comprehensive overview of this literature. A fundamental result of Lovász and Szegedy [139] establishes that if a sequence $\{G_n\}$ of dense graphs[5] has the property that for every fixed subgraph H the density[6] of copies of H in G_n converges to a limit, then there is a well-defined object (i.e., a graphon) by which all of

[5] Here, 'dense' means that, for $G_n = (V(G_n), E(G_n))$, where $|V(G_n)| = n$, we have $|E(G_n)| = \Omega(n^2)$ (i.e., the number of edges is lower-bounded by cn^2 for some constant $c > 0$).

[6] Formally, we refer to the so-called homomorphism density $t(H,G_n) = \hom(H,G_n)/|V(G_n)|^{|V(H)|}$, where $\hom(H,G_n)$ counts the number of edge-preserving maps $V(H) \rightarrow V(G_n)$.

these limits are defined. In light of the classical characterization in probability theory of a distribution function through its moments, the above result suggests an alternative sense in which a graphon may be thought of as a distribution on (infinite) random graphs. From a mathematical perspective, the theory of graph limits allows tools of analysis to be brought to bear in a context that would otherwise seem to be (on the surface, at least) purely combinatorial. The connection between exchangeability in random graphs and graph limits appears to have been independently noted and exposited upon by Diaconis and Janson [62] and by Austin [19].

2.3 Things We Demand of Our Models: Progress and Challenges

The classes of latent network models described in the previous section arguably are relatively sophisticated – at least in comparison to, say, classical linear models as encountered in a standard graduate course. On the other hand, in the case of linear models, we can say with confidence that the vast majority of what one might want to know and have of these models is indeed known and had. In contrast, we are still in the comparatively early stages of meeting the analogous demands for network models in general. But in many ways we are furthest along towards this goal in the case of latent network models. In this section, we look briefly at both progress and challenges in this direction. Motivated in part by the comment of Robins and Morris quoted at the start of this chapter, we consider four topics: plausibility, estimation, consistency and related properties, and goodness of fit.

2.3.1 *Plausibility*

To what extent can the latent network models we have seen be expected to be appropriate for modeling real-world networks? That is, how plausible are these models?

Consider the stochastic block model defined in (2.1)–(2.2), which, we recall, is essentially a mixture of classical random graph models. While the latter are understood to be too simplistic for most real-world data, mixtures of such turn out to be surprisingly flexible. For example, Figure 2.2 shows a visualization of a subnetwork of French political

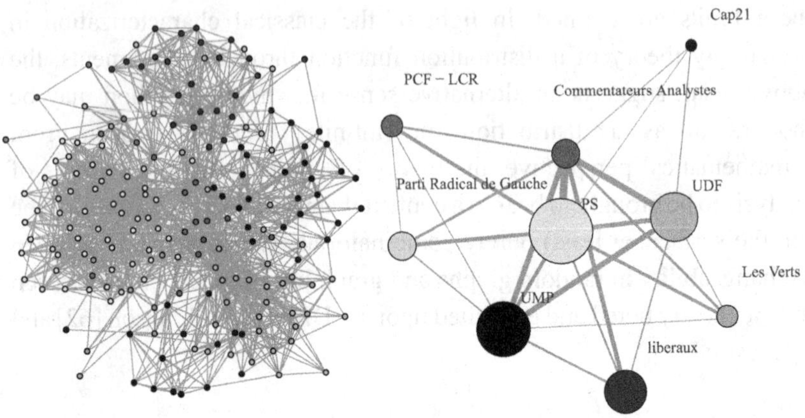

Figure 2.2 Visualizations of the French political blog network. *Left:* Original network, with colors indicating correspondence with one of nine political parties. *Right:* The same network, coarsened to the level of political party.

blogs, extracted from a snapshot of over 1100 such blogs on a single day in October 2006 and classified by the 'Observatoire Présidentielle' project as to political affiliation.[7] The network consists of 192 blogs linked by 1431 edges, the latter indicating that at least one of the two blogs referenced the other. Vertices are colored to indicate affiliation of each blog with one of nine French political parties. Visually, there is some evidence to suggest a mixing of smaller subgraphs, roughly corresponding to political party affiliation. And certainly it is not unreasonable to expect from the context of these data (i.e., political blogs in the run-up to the French 2007 presidential election) that subsets of blogs reference other subsets at differential rates that are driven at least in part by party affiliation. At first glance, therefore, it seems that the stochastic block model may not be an unreasonable approximation to reality in this case.

This conclusion is further supported in coarsening this network by party affiliation, as also shown in Figure 2.2. In this visualization, the size of the groups defined by political parties in the original network, and the numbers of edges between those groups, are reflected in vertex size and edge thickness, respectively. The relationships within and

[7] Original source: http://observatoire-presidentielle.fr/. The subnetwork used here is part of the **mixer** package in R. Note that the inherent directionality of blogs is ignored in these data, as the network graph is undirected.

among political parties that were only suggested by the original visualization are now quite evident.

While the above example is heuristic – wherein we suggest vaguely that a given graph can be represented to some reasonable extent by another of a certain type – the notion of an approximation of one graph by another actually can be made formal and precise. And a celebrated result from mathematics – the Szemerédi regularity lemma – can be used to provide some sense of the accuracy to which an arbitrary graph G can be approximated by a stochastic block model. See, for example, [138, Ch 9] for a comprehensive summary of both the original lemma and variations thereof, both weaker and stronger. Here we present a weaker version sufficient for our purposes, due to Frieze and Kannan [84], as modified slightly by Borgs and colleagues [35].

To discuss approximation accuracy we require a distance between graphs.[8] For two graphs G and G' with the same number of vertices N_v, and possible edge weights $0 \leq w_{ij} \leq 1$, the *cut distance* is defined as

$$d_{\square}\left(G, G'\right) = \frac{1}{N_v^2} \max_{S,T \subseteq \{1,\dots,N_v\}} |e_G(S,T) - e_{G'}(S,T)|, \tag{2.6}$$

where

$$e_G(S,T) = \sum_{i \in S} \sum_{j \in T} w_{ij}(G)$$

(and, analogously, $e_{G'}(S,T)$) is the sum of the weights of edges in G that connect a vertex in S with another in T. In the case that G is unweighted, as assumed throughout most of this chapter, this sum is simply the number of edges between vertices in S and vertices in T. It is straightforward to show that the quantity $d_{\square}(,)$ is a formal metric.[9]

[8] The definition of distances between graphs, and the study of their behavior and properties, is a topic of some nontrivial interest in its own right. See, for example, [138, Ch 8] for an introduction to this topic. The topic becomes particularly interesting as one moves from dense graphs to nondense graphs (i.e., so-called sparse graphs). See [33] for seminal work on this aspect.

[9] Furthermore, sometimes useful is the fact that it may be expressed equivalently in terms of the adjacency matrices for G and G', say A and A'. That is, $d_{\square}\left(A, A'\right) = ||A - A'||_{\square}$, where

$$||A||_{\square} = \frac{1}{N_v^2} \max_{S,T \subseteq \{1,\dots,N_v\}} \left| \sum_{i \in S, j \in T} A_{ij} \right|.$$

Now let $P = \{V_1, \ldots, V_Q\}$ be a partition of the vertices V of a graph G into Q subsets. Define G_P to be the complete graph on N_v vertices, with edge weights $w_{ij}(G_P) = e_G(V_q, V_r)/|V_q||V_r|$ between any vertices $i \in V_q$ and $j \in V_r$. This object may be thought of as the expectation of a block model approximation to G, with Q classes, where the probability of an edge between a pair of vertices i and j is just the weight $w_{ij}(G_P)$.

Our variant of the regularity lemma [35, Lem 3.1] then says that, for any $\in > 0$ and every graph G, there exists a partition P of the vertices V of G into $Q \leq 2^{2/\in^2}$ classes such that $d_\square(G, G_P) \leq \in$. In other words, any arbitary graph G can be approximated to arbitrary accuracy $\in$ by (the expectation of) a block model if we are willing to use up to $2^{2/\in^2}$ classes to do so. We have, therefore, some not-unreasonable justification to support the often-made assertion that stochastic block models can approximate arbitrary graphs arbitrarily well. Unfortunately, some caution clearly is appropriate here, since the upper bound provided on Q can be prohibitively large. And work in the mathematics literature (e.g., [94]) has demonstrated graphs for which the upper bounds produced by such regularity lemmas are not far off.

Now consider the graphon-based model defined in (2.3)–(2.4). On the surface, this model would appear to be extremely flexible, since the definition of the graphon f is rather generic. However, in fact, this model is appropriate only for modeling dense networks, since, if a random graph G of N_v vertices is exchangeable,[10] it is either dense or empty. This latter claim follows by noting that the expected proportion of $N_v(N_v - 1)/2$ possible edges present in G is simply

$$(1/2)\int_{[0,1]^2} f(x,y)\,dxdy.$$

Since this quantity is independent of N_v, the expected number of edges in G is $O(N_v^2)$. Models for sparse graphs can be obtained in this setting by introducing a scaling factor that forces the probability of an edge between two vertices i and j to tend to zero at some prespecified rate (e.g., [24, 214]). Alternatively, recent work by Caron and Fox [41] fundamentally changes the underlying graphon-based model to define a class of models that allows both dense and sparse random graphs.

[10] Formally, if this random graph is the restriction to vertices $\{1, \ldots, N_v\}$ of an infinite random graph whose adjacency matrix is exchangeable.

The above arguments suggest that the verdict is mixed on the plausibility of both the basic stochastic block model and the graph-based model. From a purely practical point of view, however, the main barrier impeding wider relevance of these models is the central property of vertex exchangeability. Models with this property are most natural for unlabeled graphs (i.e., graphs for which no distinction is made among the vertices, other than that rendered by their pattern of adjacencies). In reality, however, for most networks observed in practice the vertices are labeled – and the labels typically matter. Individuals in a social network will differ in personality and characteristics. Proteins in a network of protein–protein interactions have each evolved to play specialized roles in very different biochemical processes within the life of a cell. And engineered networks, like the Internet or energy networks, generally have a variety of different devices that each serve specific purposes (e.g., such as at the various levels of the router-level Internet hierarchy).

Incorporating vertex characteristics (when available) as covariates is a natural solution here. To date, however, this approach appears to have been explored only to a limited extent (e.g., [7, 190, 191]). Alternatively, there is also recent work on exploring the potential of other forms of exchangeability in network modeling, such as edge exchangeability [58]. Similarly, the use of covariate information for edges has been proposed as well (e.g., [147]).

2.3.2 Inference

We turn now to the question of inference in our two classes of latent network models. Research on this question has been highly active in recent years, yielding already too many proposed methods of inference to summarize here. Instead, we focus on presenting only a select subset of methods felt to be both representative and – collectively – amenable to concise exposition. We will see that for both parametric and nonparametric models, there are close similarities in both the inferential challenges faced and the solutions proposed.

Inference in the Stochastic Block Model

For the class of stochastic block models, recall that these models are defined up to the unknown parameters $\alpha = \{\alpha_q\}_{q=1}^{Q}$ and $\pi = \{\pi_{qr}\}_{1 \le q,r \le Q}$. And paralleling these parameters are the random variables $\mathbf{Z} = \{\{Z_{iq}\}_{q=1}^{Q}\}_{i \in V}$, for $Z_{iq} = I_{i \in C_q}$, and the $N_v \times N_v$ random matrix $\mathbf{Y} = [Y_{ij}]$, where $Y_{ij} = I_{\{i,j\} \in E}$. Since only $\mathbf{Y}$ is observed, and not $\mathbf{Z}$, inference in the context of these models focuses on both estimation of the parameters and assignment of class labels, based on the observed adjacency matrix, $\mathbf{Y} = \mathbf{y}$. In fact, frequently the classification is of primary interest, and the estimation secondary. Note that here we assume the number of classes Q to be known. This assumption, of course, is rarely true in practice. We will return to the important challenge of selecting the number of classes in Section 2.3.4.

Given the probabilistic nature of the stochastic block model, a maximum likelihood approach to estimation is natural here. Let $\theta = (\alpha, \pi)$ denote the full set of parameters to be estimated. In the event that we somehow knew the class membership information, say $\mathbf{Z} = \mathbf{z}$, the (complete data) log-likelihood would take the form

$$\ell(\mathbf{y}, \mathbf{z}; \theta) = \sum_i \sum_q z_{iq} \log \alpha_q + \frac{1}{2} \sum_{i \neq j} \sum_{q \neq r} z_{iq} z_{jr} \log b(y_{ij}; \pi_{qr}), \qquad (2.7)$$

where $b(y; \pi) = \pi^y (1 - \pi)^{1-y}$. Unfortunately, direct evaluation of the observed data likelihood, via the expression $\ell(\mathbf{y}; \theta) = \log\{\sum_{\mathbf{z}} \exp[\ell(\mathbf{y}, \mathbf{z}; \theta)]\}$, is intractable in any problems of real interest.

The formulation of our model in terms of mixtures suggests the use of the expectation-maximization (EM) algorithm [61] as a natural alternative here, alternating between producing estimates of the conditional expectations $\mathbb{E}[Z_{iq}|\mathbf{Y}] = \mathbb{P}(Z_{iq} = 1|\mathbf{Y})$, on the one hand, and of the parameters in θ, on the other. This approach was developed by Snijders and Nowicki [179], for the case of $Q = 2$ classes. But implementation involves evaluation of the conditional distributions $\mathbb{P}(\mathbf{Z}|\mathbf{Y})$, which is itself prohibitively expensive in most applications. The same authors also explore posterior-based inference in a Bayesian variant of the model, through the use of Gibbs sampling, but this approach too scales poorly to large networks.

Two popular alternatives currently, with significantly better computational efficiency, are the use of (i) variational maximum likelihood and (ii) maximum profile likelihood, both of which replace the likelihood function $\ell(\mathbf{y}; \theta)$ with an alternative objective function and employ alternating maximization strategies. In addition, both approaches have been shown to yield consistent assignment of vertex class and estimation of the parameters θ, under appropriate conditions (as we detail below, in Section 2.3.3).

The variational approach was developed by Daudin and colleagues [59]. Following the standard paradigm for such variational approaches [120], the idea is to optimize a lower bound of the likelihood $\ell(\mathbf{y};\theta)$, of the form

$$J(R_{\mathbf{y}};\theta) = \ell(\mathbf{y};\theta) - \mathrm{KL}[R_{\mathbf{y}}(\cdot), \mathbb{P}(\cdot|\mathbf{y})], \tag{2.8}$$

where KL denotes the Kullback–Leibler divergence, $\mathbb{P}(\mathbf{Z}|\mathbf{y})$ is the conditional distribution of $\mathbf{Z}$ given $\mathbf{Y} = \mathbf{y}$, and $R_{\mathbf{y}}(\mathbf{Z})$ is the variational approximation to that distribution. Using a mean-field approximation to $\mathbb{P}(\mathbf{Z}|\mathbf{Y})$, $R_{\mathbf{y}}$ is restricted to the form

$$R_{\mathbf{y}}(\mathbf{z}) = \prod_{i=1}^{N_v} h(\mathbf{z}_i;\tau_i), \tag{2.9}$$

with $h(\mathbf{z};\tau)$ corresponding to the probability mass function of a multinomial distribution, with one event and probability vector τ. Optimization of (2.8) is then pursued by alternating between the following pair of optimizations:

$$\theta^{(t+1)} = \arg\max_{\theta} J(R_{\mathbf{y}};\{\tau_i^{(t)}\},\theta),$$

$$\{\tau_i^{(t+1)}\} = \arg\max_{\{\tau_i\}} J(R_{\mathbf{y}};\{\tau_i\},\theta^{(t+1)}). \tag{2.10}$$

The first of these two optimization problems yields closed-form solutions for the elements of θ, while the solution to the second can be shown to satisfy a fixed-point relation of relatively simple form. See [59], where it is also shown that the value of J in (2.8) at the solutions for each iteration t is necessarily nondecreasing. This approach has been implemented in the R package **mixer**.

The profile-likelihood approach has been advanced by, for example, Bickel and Chen [24] and Choi and colleagues [47]. Here we summarize this approach using notation in the spirit of the latter. Recall from the definition in (2.1)–(2.2) that, conditional on the class labels $\mathbf{z}$, the elements of the adjacency matrix $\mathbf{Y}$ follow independent Bernoulli distributions, with parameters π_{qr}. We express the corresponding conditional likelihood as

$$\ell(\mathbf{y};\mathbf{z},\pi) = \sum_{i<j} y_{ij}\log \pi_{\mathbf{z}_i,\mathbf{z}_j} + (1-y_{ij})\log(1-\pi_{\mathbf{z}_i,\mathbf{z}_j}). \tag{2.11}$$

Under the profile-likelihood approach, the $Q \times Q$ parameter matrix π is treated as a nuisance parameter, with the assignment of class labels to the vertices taken as the primary goal. This motivates definition of the estimator

$$\hat{\mathbf{z}} = \arg\max_{\mathbf{z}} \ \ell\Big(\mathbf{y}; \mathbf{z}, \hat{\pi}(\mathbf{z})\Big), \tag{2.12}$$

where

$$\hat{\pi}(\mathbf{z}) = \arg\max_{\pi} \ell(\mathbf{y}; \mathbf{z}, \pi). \tag{2.13}$$

Note that $\mathbf{z}$ is treated as fixed in the optimization of (2.13), and that the solution in that case is simply given by

$$\hat{\pi}_{qr}(\mathbf{z}) = \frac{1}{n_{qr}} \sum_{i<j} y_{ij} I\big(z_{iq} = 1, z_{jr} = 1\big), \tag{2.14}$$

for each q and r, where n_{qr} is the maximum number of possible edges between classes q and r. Similar to the variational approach described just above, an iterative numerical strategy is necessary for implementation. In [47], Markov chain Monte Carlo is used to explore the space of $\mathbf{z}$, while holding π constant.

Inference of Graphons

In the nonparametric model of (2.3)–(2.4), the task of inference is focused on estimation of the graphon, f. As noted earlier, this problem is closely related to that of nonparametric regression. Recalling that a graphon can be interpreted as a distribution on the space of (infinite) graphs, intuitively it would seem that methods based on histograms and related smoothing might be feasible. And, indeed, a collection of proposed methods has arisen around this idea. A certain version of this approach was pioneered by Kallenberg [121], well over a decade ago. The approach has independently been developed further only in the past few years, in work such as Wolfe and Olhede [205], Chan and Airoldi [43], and Gao and colleagues [87].

Histogram estimators in this context are defined in terms of averages of the entries of the adjacency matrix $\mathbf{Y}$ over groupings of vertices. Specifically, as blockwise-constant functions $\hat{f}$ on $[0, 1]^2$, with heights defined in appropriate correspondence with such averages. At the risk of some abuse of notation, let $\mathcal{Z}_{N_v,Q}$ be the set of all mappings z from $\{1,\ldots,N_v\}$ to $\{1,\ldots,Q\}$. And denote by $z^{-1}(q)$ the set of all vertices assigned the label q by the mapping z, for $q = 1,\ldots,Q$. Then define the corresponding blockwise averages of **Y**, with respect to the grouping z, as

$$\overline{Y}_{qr}(z) = \frac{1}{|z^{-1}(q)||z^{-1}(r)|} \sum_{i \in z^{-1}(q)} \sum_{j \in z^{-1}(r)} y_{ij}. \tag{2.15}$$

With an appropriate grouping, under reasonable conditions on f, it does not seem unreasonable to hope that a histogram estimator $\hat{f}$ defined through these quantities will perform well.

The key challenge here, however, is that we do not know *a priori* the most appropriate choice of grouping z to use. What makes the problem in this context unique – say, in comparison to nonparametric regression – is that we lack the appropriate ordering of the entries in **Y**, relative to the underlying latent variables $U_1, \ldots, U_{N_v}$, that would render this problem fairly straightforward in mapping to the unit square $[0,\ 1]^2$. The illustration in Figure 2.1, which explicitly used this ordering for purposes of visualization, demonstrates how this knowledge would leave us with a problem effectively equivalent to nonparametric regression. Depending on our goals, it can be sufficient – from a theoretical perspective – to only identify an appropriate grouping. For practical implementation, however, it seems a decision must be made on ordering as well.

The typical approach to grouping makes explicit use of the Szemerédi regularity lemma. Specifically, the argument is made that an appropriate choice of grouping is that corresponding to a stochastic block model approximation of the graphon f with sufficiently small approximation error. Here, 'sufficiently small' is meant in comparison to the estimation error corresponding to the estimator defined with respect to this grouping, with the understanding that a good overall estimator will balance approximation and estimation errors. Following this reasoning, methods for learning vertex class labels in stochastic block models, as discussed in the previous subsection, can then be brought to bear to infer the values of **z** corresponding to this best approximation. The inferred **z** then proscribes our choice of mapping z used in (2.15) above. Wolfe and Olhede [205], for example, recommend using the method of maximum profile likelihood described previously, and are able to characterize the mean-squared error rate of convergence of the overall resulting estimator, as we will see shortly, in Section 2.3.3. Gao and colleagues [87] do similarly for an estimator analogous to that of

Wolfe and Olhede that replaces the likelihood by a least-squares criterion.

To address the problem of ordering, additional restrictions are needed on the graphon f itself, in order to make the problem well-posed. That is, the identifiability problem discussed earlier must be addressed. For example, Chan and Airoldi [43] adopt the strict monotonicity criterion and focus on estimation of the canonical graphon, $\widetilde{f}$. The condition on the graphon degree function in (2.5) motivates these authors to propose ordering the vertices according to their observed degree. The overall resulting estimation algorithm then consists of two steps – (i) ordering based on observed degree, followed by (ii) a smoothed version of histogram estimation – and has both demonstrable theoretical properties and a computationally efficient implementation. In particular, the complexity of the algorithm is $O(N_v \log N_v + Q^2 \log Q^2)$ multiplications and $O(N_v^2)$ additions, where Q is the number of bins used.

Alternatively, a greedy alternative requiring less stringent assumptions than monotonicity has been proposed to address the problem of ordering. Based on the notion of what has been called 'neighborhood smoothing' by Zhang and colleagues [213], this approach too results in methods that are both computationally feasible and allow for establishing consistency and related theoretical properties. Airoldi and colleagues [8] employ the same basic idea in earlier work, under a problem variant that assumes a sample of multiple networks from the same graphon-based model.

2.3.3 Consistency and Other Properties

For the inferential procedures just described, we would like to know that, at a minimum, appropriate notions of consistency have been established. Such is indeed the case and, moreover, work on limiting distributions, rates of convergence, and related has been developing quickly, most of it in just the last few years. Importantly, note that for these results the notion of 'increasing sample size' generally is meant in the sense of a 'large' network – formally, by letting N_v tend to infinity. We summarize here some results of this nature.

Properties for Stochastic Block Models

Recall that for stochastic block models, the primary interest typically is in accurate assignment of class labels. The earliest consistency results relating to this topic appear to be due to Snijders and Nowicki [179], in the case of $Q = 2$ classes, for a method that assigned class membership based on the largest gap in the empirical degree sequence.

Among the methods of inference described above, the first general results were shown in the case of the profile-likelihood approach, by Bickel and Chen [24]. Specifically, they show, under appropriate conditions, that[11]

$$\mathbb{P}\left(\hat{\mathbf{Z}} = \mathbf{z} \middle| \mathbf{Z} = \mathbf{z}\right) \to 1, \tag{2.16}$$

as $N_v \to \infty$, where $\hat{\mathbf{Z}}$ is defined by (2.12). The conditions used in the proof of this result are of three types. The first is that the elements of α are sufficiently far from 0 and 1, while the rows/columns of π are sufficiently distinct. These amount to an assumption that there are indeed Q separate and nontrivial communities. Second, there are certain technical conditions, regarding regularity of relevant functions and well-posedness of the problem in the limit of large N_v. Third, and perhaps of most interest from a practical perspective, is the assumption that the expected average degree grows faster than $\log N_v$. Rephrased – by exploiting the fact that the average degree of an undirected graph is equal to $2N_e/N_v$, where N_e is the number of edges – this means that N_e must grow faster than $N_v \log N_v$. This latter condition typically is interpreted as saying that the result pertains to dense networks, rather than sparse networks, as are more commonly observed in practice.

Note too that this result is for the case where the number of communities Q is kept fixed as the network grows. A result similar to that above has been established by Choi and colleagues [47], but for the case where Q is allowed to grow as well. Using a different proof technique, these authors show that

$$\left|\left\{i : \hat{\mathbf{z}}_i \neq \mathbf{z}_i\right\}\right| = o_P(N_v). \tag{2.17}$$

That is, the proportion of mislabeled vertices tends to zero in probability. Again, however, the assumption of dense networks is employed, in that N_e is assumed in this case to grow faster than $N_v(\log N_v)^{3+\delta}$, for some $\delta > 0$. But the number of communities Q is allowed to grow as $O(N_v^{1/2})$.

The results (2.16) and (2.17) effectively say that, asymptotically, we can recover the label vectors $\mathbf{z}_i$ without error. In the case of fixed Q, this

[11] Here, '$\hat{\mathbf{Z}} = \mathbf{z}$' is to be interpreted as 'equal up to permutation of labels.'

fact in turn allows for application to the estimates $\hat{\alpha}(\hat{\mathbf{z}})$ and $\hat{\pi}(\hat{\mathbf{z}})$ of classical results on convergence of maximum likelihood estimates. Here the latter is as defined in (2.14), and the former is the analogous conditional maximum likelihood estimate for α. Accordingly, we obtain, for example, that

$$\sqrt{N_v}\left(\hat{\alpha}(\hat{\mathbf{z}}) - \alpha\right) \rightarrow N(0, \mathrm{Diag}(\alpha) - \alpha\alpha^T). \tag{2.18}$$

A related result (but more complicated) can be stated for convergence of a suitably normalized version of $\hat{\pi}(\hat{\mathbf{z}})$ to π. See [24, Cor 1].

For the variational method in (2.10), consistency has been shown for the resulting parameter estimates by Celisse and colleagues [42]. Employing conditions similar in spirit to those in [24], they show that the distances $d(\hat{\alpha}, \alpha)$ and $d(\hat{\pi}, \pi)$ between estimators and estimands converge to zero in probability, as $N_v \rightarrow \infty$, for arbitrary choices of metrics $d(\cdot, \cdot)$ on the corresponding parameter spaces. In addition, these authors also show consistency of the formal maximum likelihood estimates, but given the computational intractability of these estimates, this result is admittedly of primarily theoretical interest. See also [27] for follow-up work in this direction.

All of the above results are focused on characterizing a given method of inference through the extent to which individual estimators $\hat{\mathbf{z}}$, $\hat{\alpha}$, and $\hat{\pi}$ approach their respective targets, as the number of vertices grows. In the literature there are also characterization results in terms of the accuracy with which the expected adjacency matrix can be estimated. That is, the $N_v \times N_v$ matrix of entries $\mathbb{E}[Y_{ij}] = \pi_{\mathbf{z}_i \mathbf{z}_j}$. In these settings, a mean-squared-error criterion typically is used, that is

$$\mathbb{E}\left[\frac{1}{N_v^2}\sum_{i,j}\left(\hat{\pi}_{\hat{\mathbf{z}}_i,\hat{\mathbf{z}}_j} - \pi_{\mathbf{z}_i,\mathbf{z}_j}\right)^2\right], \tag{2.19}$$

and the goal is to characterize the corresponding rate of convergence. Following upon earlier partial results in this area, such as [45], Gao and colleagues [87] have recently established that a least-squares analogue of the maximum profile-likelihood estimator will have mean-squared error no worse than order

$$\frac{Q^2}{N_v^2} + \frac{\log Q}{N_v}, \tag{2.20}$$

for arbitrary (but fixed) Q. Furthermore, this rate cannot be improved upon (i.e., it is the minimax rate).

The expression in (2.20) is particularly interesting. The first part derives from the estimation part of the problem, wherein we seek to estimate $Q(Q+1)/2 \asymp Q^2$ unknown parameters with $N_v(N_v+1)/2 \asymp N_v^2$ observations. The second part derives from the fact that we must first assign Q class labels to the N_v vertices. Note that, if Q behaves asymptotically like a power of N_v, say $Q \asymp N_v^\delta$, where $\delta \in [0,1]$, then substitution yields four different regimes into which the rate of convergence may fall. Specifically, under this parameterization

$$\frac{Q^2}{N_v^2} + \frac{\log Q}{N_v} \asymp \begin{cases} N_v^{-2}, & \text{if } Q = 1, \\ N_v^{-1}, & \text{if } \delta = 0, Q \geq 2, \\ N_v^{-1} \log N_v, & \text{if } \delta \in (0, 1/2], \\ N_v^{-2(1-\delta)}, & \text{if } \delta \in (1/2, 1]. \end{cases} \tag{2.21}$$

As Q grows, the rate becomes increasingly slower, beginning with a sharp change at $Q = 2$, and behaving like $N_v^{-1} \log N_v$ while $Q = O(N_v^{1/2})$.

Properties for Graphon Estimation

As we have noted, work on statistical estimation of graphons is arguably still in its infancy. So too, therefore, is the corresponding work on properties of graphon estimators. The problem of interest is to show convergence of a given estimator to an appropriate target graphon in some well-defined sense. (Recall the challenges inherent to this problem, due to its particular flavor of identifiability issues.) The early work by Kallenberg [121] uses as a criterion notions of convergence in distribution of random processes, and establishes such for histogram-like estimators based on grids, for random exchangeable arrays of (fixed) arbitrary dimension. The more recent work has used mean-squared error as a criterion – either discrete variants analogous to that in (2.19) or continuous analogues thereof. Additionally, the graphon of interest, f, is assumed to have certain smoothness properties, as a function on the unit square $[0,1]^2$.

Chan and Airoldi [43], focusing on estimation of the canonical graphon $\widetilde{f}$, place a Lipschitz condition on both $\widetilde{f}$ and the corresponding degree function, $\widetilde{g}$. Additionally, they make a sparsity assumption for the gradient of $\widetilde{f}$. Then, the histogram-based estimator produced by their proposed algorithm has mean-squared error that behaves like $O\left((\log N_v)/N_v\right)$, and hence tends to zero at a rate slightly worse than N_v^{-1}. In particular, it is a consistent estimator.

Similarly, Gao and colleagues [87] assume that the graphon f of interest comes from a Hölder class with smoothness $\alpha > 0$. They show that for their least-squares-based estimator, the corresponding mean-squared error behaves either like $O\left(N_v^{-2\alpha/(\alpha+1)}\right)$, when $0 < \alpha < 1$, or like $O\left((\log N_v)/N_v\right)$, when $\alpha \geq 1$. In addition, in this case too this behavior is shown to be minimax.

Finally, Wolfe and Olhede [205] provide rather general bounds on the mean-squared error, which they then use to study several classes of graphons f, focusing in particular on cases of both dense and sparse networks. See [205] for details. See too [157], by the same authors, where a practical variant of this methodology is proposed, with the goal of producing a histogram-like visualization of the underlying graphon. Analogies to histogram bin-width selection are exploited, and use is made of vertex covariate information to address the problem of ordering the vertices along the axes of the unit square.

2.3.4 Goodness of Fit

The results on consistency and related properties of estimators in the latent network models we have considered are encouraging. But our earlier discussion of the plausibility of such models for real-world networks suggests that goodness of fit must be another key part of the arsenal of statistical concepts and tools accompanying these models. The amount of work on this topic to date has been comparatively limited and is confined almost entirely to the context of stochastic block models. Moreover, currently, most methods proposed for assessing goodness of fit in network modeling are computational in nature and rely heavily on visualization. However, this state of affairs is beginning to change, as progress on the theoretical underpinnings is starting to be made.

Computational Assessment

Figure 2.3 shows an example of the type of visualization-based goodness-of-fit assessment available for stochastic block models. The model was fit using the **mixer** package in R, using a Bayesian version of the variational

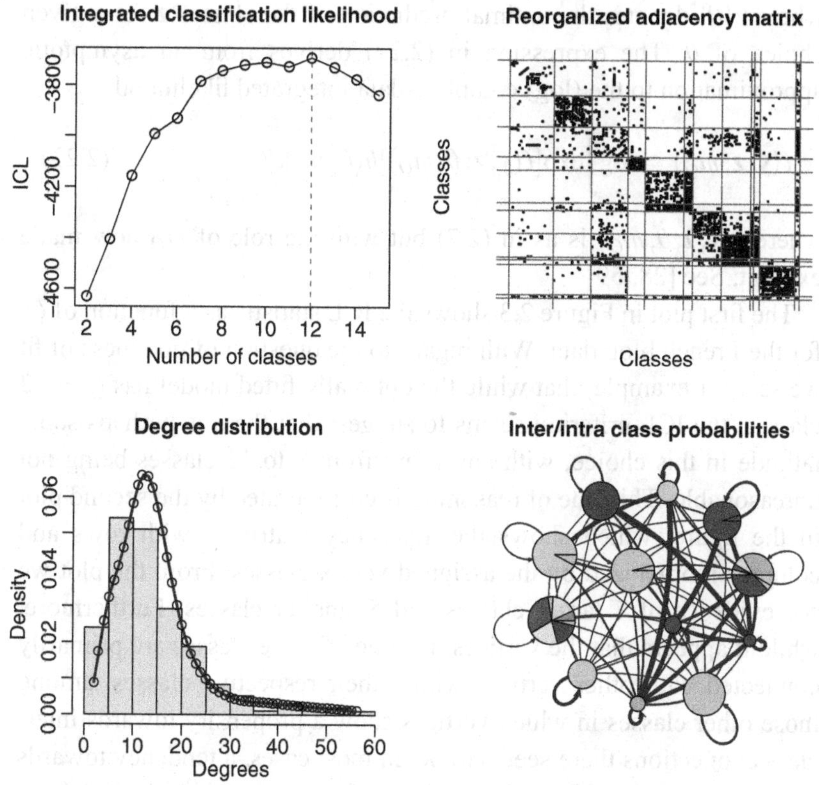

Figure 2.3 Various plots summarizing the goodness of fit for the stochastic block model analysis of the French political blog network.

method introduced earlier in this chapter. The number of classes Q is chosen using principles of model selection, analogous to the Bayesian information criterion (BIC) method in traditional regression analysis.

Specifically, the integrated classification likelihood (ICL) criterion is applied. Originally developed by Biernacki and colleagues [28] in the context of clustering with Gaussian mixture models, it was adapted by Daudin and colleagues [59] to the present context. Let m_Q denote a stochastic block model of Q classes, and let $h(\cdot|m_Q)$ be a prior distribution on the parameters $\theta = (\alpha, \pi)$, given that the model contains Q classes. The ICL statistic proposed in [59] takes the form

$$\mathrm{ICL}(m_Q) = \max_{\theta} \ell\left(\mathbf{y}, \hat{\mathbf{z}}(\theta); \theta, m_Q\right) - \frac{Q(Q+1)}{4} \log \frac{N_v(N_v - 1)}{2} - \frac{Q-1}{2} \log N_v, \qquad (2.22)$$

where $\hat{\mathbf{z}}(\theta)$ denotes the optimal prediction of the classes $\mathbf{z}$ for a given choice of θ. The expression in (2.22) derives from an asymptotic approximation to the (log) complete-data integrated likelihood

$$\ell(\mathbf{y}, \mathbf{z}; m_Q) = \log \int \exp[\ell(\mathbf{y}, \mathbf{z}; \theta, m_Q)] h(\theta | m_Q) d\theta, \qquad (2.23)$$

where $\ell(\mathbf{y}, \mathbf{z}; \theta, m_Q)$ is as in (2.7) but with the role of m_Q now made explicit. See [28, 59].

The first plot in Figure 2.3 shows the ICL statistic as a function of Q, for the French blog data. With regard to the question of goodness of fit we see, for example, that while the optimally fitted model has $Q = 12$ classes, the ICL criterion seems to suggest that there is perhaps some latitude in this choice, with anywhere from 8 to 12 classes being not unreasonable. This line of reasoning is corroborated by the second plot in the figure, which shows the adjacency matrix $\mathbf{y}$, with rows and columns reorganized by the assigned vertex classes. From the plot we see evidence of 7 larger classes and 5 smaller classes. Furthermore, while it appears that the vertices in some of these classes are primarily connected with other vertices within their respective classes, among those other classes in which vertices show a propensity towards inter-class connections there seems to be, in most cases, a tendency towards connecting selectively with vertices of only certain other classes. In other words, the potential for merging some of the putative 12 classes seems compelling.

Shown in the third plot is an example of a popular approach to assessing goodness of fit in network modeling quite generally. There we see plotted (i) the degree distribution of the observed network (yellow histogram) and (ii) the expected degree distribution under the fitted stochastic block model (blue curve). The latter is obtained through Monte Carlo simulation. Of course, any summary statistic(s) of the network can in principle be used. Nevertheless, the degree distribution is a natural starting place with such an approach. Examining the plot, it appears that the distribution corresponding to the fitted stochastic block model is able to describe the observed degree distribution reasonably well, although the body of the fitted distribution is arguably shifted to the right somewhat of the observed distribution.

Finally, for these particular data, we have an existing set of class labels (although not necessarily the 'truth') and can use those as well in assessing goodness of fit. Specifically, we can consider to what extent the model-based vertex class assignments match the grouping of these blogs according to their political party status. This comparison is summarized in the last plot in Figure 2.3. Here the circles, corresponding to the 12 vertex classes, and proportional in size to the number of blogs assigned to each class, are further broken down according to the relative proportion of political parties to which the blogs correspond, displayed in the form of pie charts. Connecting the circles are edges drawn with a thickness in proportion to the estimated probability that blogs in the two respective groups link to each other (i.e., in proportion to the estimated π_{qr}). Note that this plot may be contrasted with the coarse-level visualization of the original French blog network in Figure 2.2.

A close examination of the pie charts yields, for example, that while the blogs in most of the 12 classes are quite homogeneous in their political party affiliations, two of the larger classes have a rather heterogeneous set of affiliations represented. In addition, two of the political parties (shown in light blue and light green) appear to be split largely between two classes, one larger and one smaller, while another (blue) appears to be mainly split among four classes, two larger and two smaller. This latter observation might suggest that the model has chosen to use too many classes. Alternatively, it could instead indicate that there is actually splintering within these political parties.

Theory Supporting Goodness of Fit

Formal goodness of fit is frequently assessed in statistical modeling through the use of principles of hypothesis testing. Results in this spirit were developed recently for stochastic block models by Lei [134]. (See [25] as well, for similar results of a more restricted nature.) There, two related problems are considered. The first problem is to decide whether or not a stochastic block model with a hypothesized number of classes, say Q_0, is appropriate for an observed network. In contrast, the second problem is to determine what number of classes Q should be used. Lei proposes a hypothesis-testing procedure to address the first problem, and then develops a sequential testing approach to address the second. Asymptotic control of Type I error and demonstration of power against alternatives within the stochastic block model class are provided, under appropriate conditions, as is – ultimately – a guarantee of consistency in estimating Q.

More formally, suppose our adjacency matrix $\mathbf{Y}$ comes from a stochastic block model in some class m_Q. Consider testing $H_0 : Q = Q_0$ against the one-sided alternative $H_1 : Q > Q_0$, for some Q_0. Given an inferred assignment of class membership $\hat{\mathbf{z}}$, using Q_0 classes, let $\hat{\pi}_{\hat{\mathbf{z}}_i,\hat{\mathbf{z}}_j}$ be the induced estimate of $\mathbb{E}[Y_{ij}]$, defined through block averaging analogous to (2.15). Then, define the (standardized) residuals from the fitted stochastic block model to be

$$\widetilde{Y}_{ij} = \begin{cases} Y_{ij} - \dfrac{\hat{\pi}_{\hat{\mathbf{z}}_i,\hat{\mathbf{z}}_j}}{\sqrt{(N_v - 1)\hat{\pi}_{\hat{\mathbf{z}}_i,\hat{\mathbf{z}}_j}(1 - \hat{\pi}_{\hat{\mathbf{z}}_i,\hat{\mathbf{z}}_j})}}, & i \neq j, \\ 0, & i = j. \end{cases} \tag{2.24}$$

Under H_0, if the estimates $\hat{\mathbf{z}}$ and $\hat{\pi}_{\hat{\mathbf{z}}_i,\hat{\mathbf{z}}_j}$ are sufficiently accurate, it can be expected that the entries of the residual matrix $\widetilde{Y}$ behave essentially like centered and scaled Bernoulli random variables. Accordingly, results of random matrix theory can be brought to bear in characterizing the behavior of $\widetilde{Y}$. And, common to that literature, is the use of eigenvalues for characterization. Hence, a natural test statistic in this context is

$$T_{N_v,Q_0} = \max\left[N_v^{2/3}\left(\lambda_1(\widetilde{Y}) - 2\right),\ N_v^{2/3}\left(-\lambda_{N_v}(\widetilde{Y}) - 2\right)\right], \tag{2.25}$$

where $\lambda_1(\cdot)$ and $\lambda_{N_v}(\cdot)$ indicate the largest and smallest eigenvalues, respectively.

The asymptotic distribution of this statistic, as $N_v \to \infty$, can be shown to be the Tracy–Widom distribution, with index 1, abbreviated TW_1. The corresponding rule for this test is then that H_0 is rejected, at level $\gamma \in (0, 1)$, if $T_{N_v,Q_0} > tw(\gamma/2)$, where $tw(\gamma/2)$ corresponds to the $\gamma/2$ upper quantile of the TW_1 distribution. Note that $\gamma/2$ is used, rather than γ, because the statistic in (2.25) effectively defines a two-sided test.

With this test in hand, a sequential testing procedure then seems to recommend itself for the estimation of Q. That is, intuitively, we successively test increasingly larger values of Q_0 until we no longer reject. Formally, we define the estimator

$$\hat{Q} = \inf\{Q_0 \geq 1 : T_{N_v,Q_0} < t_{N_v}\}, \tag{2.26}$$

for some choice of thresholds t_{N_v} varying with the number of vertices N_v.

Under appropriate conditions on the model, use of a consistent method of class membership assignment (for any fixed Q), and an assumption that the thresholds grow at an appropriate rate in N_v, Lei

[134] shows that the test of H_0 defined above has asymptotic level γ. Furthermore, a guarantee of asymptotic power is provided in the form of a lower bound on $\max(\lambda_1(\widetilde{Y}), -\lambda_{N_v}(\widetilde{Y}))$. Finally, for the sequential testing procedure and its corresponding estimator, in (2.26), consistency is shown. That is, it is shown that

$$\mathbb{P}(\hat{Q} = Q) \rightarrow 1. \tag{2.27}$$

Importantly, Q is allowed to grow with N_v in this framework.

2.4 Related Topics

The topic of statistical network modeling is quite broad. Certainly too broad to be done full justice here. The focus in this chapter, consistent with the theme of this monograph, has been on providing a relatively brief and self-contained look at how network modeling touches the foundations of statistics and *vice versa*. For that, the chosen topic of latent network models, and the various subtopics therein upon which we have elected to focus, seems most conducive to our goals. Nevertheless, there are a variety of related topics – in the area of latent network models alone, much less network modeling in general – that deserve mention. Many of these represent relatively open and likely fertile areas for further research. A brief sketch of such topics is provided below. For a more comprehensive survey of the literature on stochastic block models, see [149]. Also see [39] for a review of the closely related literature on latent block models, with connections with stochastic block models highlighted throughout.

The notion of stochastic block models introduced here is the original and most basic of what is in fact a larger spectrum of such models. We focused upon only this particular slice of that spectrum because it is for this that statistics has, as a field, accumulated the most knowledge and understanding to date, and, additionally, because it represents the most immediate point of intersection with the more recent topic of graphon-based modeling. Extensions of the basic stochastic block model include (i) adaptation to networks with directed edges (i.e., direct graphs or digraphs; e.g., [155, 203]); (ii) relaxation of the requirement for strict membership in classes to mixed membership (e.g., [6, 130]); (iii) parameterizations that correct for the type of heterogeneous degree distributions commonly encountered in practice (e.g., [123, 214]);

(iv) the incorporation of edge weights (i.e., for so-called valued graphs; e.g., [5, 116, 147]); (v) embedding within a hierarchical structure (e.g., [53, 105]); and (vi) dynamically changing networks (e.g., [65, 85, 206]). Most recently, there is also work looking at extensions to networks with multiple layers (i.e., so-called multi-graphs [185]).

Our focus on inference for stochastic block models was restricted to just two key approaches – maximum likelihood (including variational approximations) and maximum profile likelihood. Additionally, proposals have been made for inference based on principles of methods of moments (e.g., [13, 26, 82]) and of pseudo-likelihood (e.g., [13, 14]). These methods are for inference of class labels or model parameters or both, and, in many cases, convergence results are provided. We note too that, beyond that provided by asymptotic distribution theory established in the literature, there has been additional work in the direction of confidence sets for network parameters [7], which in turn has been proposed for use in assessing goodness of fit.

More generally, there is a substantial literature on the inference of class labels alone for stochastic block models. And frequently the proposed approaches are not inherently limited to the context of stochastic block models – nor are they necessarily model-based at all. A prime example is that of spectral clustering. Representing a general class of methods for the so-called community detection problem – of which the class labeling problem in stochastic block models is a more precisely defined, model-based variant – spectral clustering invokes results and tools of spectral graph theory to motivate using the primary eigenvector(s) of certain matrix-based representations of a network to partition the network into subnetworks (a.k.a., 'communities'). See [127, Ch 4.3.3] for a brief overview of the topic. Seminal work on consistency of class-label assignments derived from spectral clustering was done by Rohe and colleagues [166], in the case where the number of classes Q is allowed to grow with the number of vertices N_v.

An important subtopic in the above literature is that of detectability, where the goal is to characterize the fundamental limits for community detection under a stochastic block model. Interestingly, various phase transitions are exhibited in this setting, wherein regimes of no recovery, 'weak' recovery, partial recovery, and exact recovery of community structure are found, depending on the parameterization of the model. Determining necessary and sufficient conditions on the parameters for each regime to obtain has been the focus of a highly active effort across multiple research communities in recent years. See the recent paper by Abbe and Sandon [1], for example, for an explicit characterization of these recovery thresholds, as well as a fairly thorough literature review.

Worth mentioning too is the importance of computational efficiency in network modeling – indeed, in network analysis quite generally. For stochastic block models, while the maximum likelihood method is understood to be computationally intractable, and methods like variational approximations and profile likelihood are improvements, the desire to push these models to the scale of the types of complex networks encountered in, say, social media and the Internet necessitates continued research on computationally efficient implementations of inferential strategies. Complicating the challenge is the fact that not only is the number of vertices N_v important, but so too can be the number of edges N_e and the number of classes Q as well. Recent work in this area allows for inference with stochastic block models in networks with hundreds of thousands of nodes (e.g., [201]) and, most recently, for a closely related class of probabilistic network models, in networks with hundreds of millions of nodes [106].

Turning now to the topic of graphons, we comment on two additional directions of note. First, while our presentation of graphon-based models was frequentist in nature, there is emerging a nontrivial literature on their use in defining priors for Bayesian network modeling. See [158], for example, for a recent survey of work in this area. Second, as already noted, there is recent interest in generalizing the Aldous–Hoover formalism associated with graphon-based statistical network models, to move beyond various of the limitations mentioned earlier in this chapter. A seminal example of such an extension in this area is the work of Caron and Fox [41], building analogues of graphons supported on the full-positive quadrant of the real plane, as opposed to the unit square (or a rescaling thereof). Their construction allows, among other things, for the possibility of both dense and sparse random graphs. See also [199]. Alternatively, Crane and Dempsey [58] have recently offered a class of models whose derivation rests on the replacement of vertex exchangeability with an analogous notion of edge exchangeability. There too it is possible to obtain sparse random graphs.

More generally, however, there remains much to be done in developing the statistical aspects of graphon-based network modeling, as we have only really just begun to scratch the surface. Clearly, the graphon has already been found to be a powerful and important theoretical device. However, additional inroads are needed as well to tap the serious potential for practical impact. For example, more attention is needed on the computational efficiency of proposed inferential strategies. Even further, the question of goodness of fit for graphon-based models appears to be untouched at present. There is a need to understand just what class(es) of graphons are suggested by real-world networks. Currently, inspired by the analogy with nonparametric function estimation, two-dimensional versions of classical function spaces (e.g., Hölder, Lipschitz, etc.) have been assumed in the literature. But such assumptions are purely for theoretical convenience. Do real-world networks support such assumptions? Conversely, what are the practical implications for real-world networks and their characteristics of assuming such classes? In particular, what is the nature of standard network characteristics of interest (e.g., degree distributions, clustering, centralities, shortest path lengths, etc.) in such models?

As an important aside, it is worth noting that another direction of departure from the graphon-based model defined in this chapter – and one already explored in some depth – is towards *more* structure rather than less. The random variables $U_1, \ldots, U_n$ in (2.3)–(2.4) can have distributions other than uniform and, in fact, can be vector-valued. This observation suggests adopting more flexibility in the distribution of the latent variables in exchange for simpler forms of the function f. A key example of this approach is the notion of random dot-product graphs, a version of network modeling based on latent positions. First introduced formally by Young and Scheinerman [210], these models assume the form of a dot product for f, between vertex-specific vectors of latent positions. These models can be more interpretable and can reduce the complexity of the non-identifiability. In addition, they are also conducive to a variety of estimation procedures based on the imposed structure of the graphon, in a manner that successfully balances practical computational implementations with theoretical tractability. See [189, 192], for example.

Recall that at the start of this chapter it was observed that there were two key approaches to network modeling in statistics. In addition to the class of latent network models that was the focus here, the class of exponential random graph models (ERGMs) deserves additional comment. Defined through notions of conditional independencies, via a graphical models perspective, these models have a long history, going back at least to the 1980s. See [164, 204], for example, for short summaries of this history, and [143] for a comprehensive treatment of the topic. The development of these models at the methodological level is extensive, as is their use in practice, facilitated by several well-established software packages for their use.[12] However, the corresponding theoretical development of these models has lagged noticeably behind. And recent progress in this direction has brought with it some reasons for concern about ERGMs – at least as traditionally employed.

Seminal work of Frank and Strauss [83] established a key version of ERGMs known as a Markov graph. In particular, they showed that a certain Markov relationship holds among entries Y_{ij} of the adjacency matrix $\mathbf{Y}$ if and only if $\mathbb{P}(\mathbf{Y})$ has a certain form. Specifically, it must have the form of an exponential family model, with endogenously defined independent variables (i.e., defined as functions of $\mathbf{Y}$) consisting of counts of certain basic subgraphs – the number of edges, the number of successively larger stars, and the number of triangles. Various practical extensions of ERGMs have traditionally had this model at their core, as it has the key and important characteristic of being defined in terms of what are considered fundamental social microstructures, hence allowing for estimation and testing of the impact of such social units on network formation. Unfortunately, however, it was discovered that such models have the undesirable property of placing nearly all of their mass on what is essentially either the empty graph (i.e., the graph without any edges) or the complete graph (i.e., the graph with all edges), depending on the model parameters. See [100, 160] for early characterizations of this degeneracy, and [46] for a definitive treatment. Corresponding to this phenomenon in turn are problems of convergence of iterative procedures for producing maximum likelihood estimates of model parameters and of goodness of fit more generally. Alternative formulations of these models have been proposed, which appear to avoid these issues (e.g., [165]), although they lack the easy interpretability of the original formulation.

The ERGM model class has proven to be particularly resistant to theoretical study, due in no small part to the nature of the normalization constant and its connection to the above issues [174]. For example, while the ERGM class is routinely thought about as essentially a variant of generalized linear models – and, indeed, treated as such in

[12] Within the R community, arguably the most popular is the **ergm** package within the **statnet** suite of tools. See [128, Ch 6.2], for a brief introduction.

applications – consistency and asymptotic normality of maximum likelihood estimates has only been established in fairly limited contexts (e.g., [46, 131]). When questions of sampling get brought into the picture, the story appears to become more complicated still [176]. Recent traction in this area appears to be gained by moving slightly away from the rigid form dictated by, for instance, the Markov assumptions in [83], such as through a focus on sufficient statistics [44] or on local dependencies [175].

On a final note, it is important to point out that the literature on network modeling in general – that is, not just from a statistical perspective – is exceedingly broad and, in many areas, quite deep. Entire major sub-literatures on the topic exist, from the perspectives of the economic and social sciences (e.g., [115]), mathematics and probability theory (e.g., [50, 67]), and statistical physics (e.g., [154]). These literatures are an excellent source not only of foundational statistical topics remaining to be revisited, but also for staying abreast of what is considered emerging in the broader network science arena. The latter currently includes, in particular, topics like dynamically evolving networks (i.e., longitudinal networks, network time series, and network point processes) and multi-networks (i.e., networks with multiple types of vertices and/or edges).

3 Sampling Networks

3.1 Introduction

Throughout the previous chapter, the implicit assumption was that we observed a network G (or, equivalently, its adjacency matrix $\mathbf{Y}$) in its entirety. Frequently, however, it is the case that relational information is observed on only a portion of a complex system being studied, and the network resulting from such measurements may be thought of as a sample from a larger underlying network. In fact, arguably sampled networks may be more the rule than the exception in practice. Yet sampling in networks, and the corresponding statistical inference problems that arise, have to date been noticeably less studied than, for example, network modeling.

Network sampling arises in a variety of contexts. For example, in social networks, while it may be possible to fully construct the friendship network among school children in a small classroom, it may be cumbersome to attempt to do so for all employees in a large corporation. Similarly, sampling is a necessary part of various measurement tasks relating to Internet networks, such as topology studies that help in 'mapping' the Internet or 'crawls' aimed at collecting web pages and their citation patterns. Finally, in the context of many biological networks, sampling plays a key role too, such as that induced by design considerations into protein–protein interaction networks constructed from affinity-binding experiments or by spatial design choices in surveys underlying predator–prey networks.

How sampling impacts a statistical network analysis would appear to depend in no small part on the type of analysis to be done. Here in this chapter we will focus on the task of characterizing the structural properties of a network. A substantial fraction of the foundational work done on this topic is due to Ove Frank and colleagues, appearing in a series of papers throughout the 1970s and 1980s. See Frank [81] for a recent overview and extensive bibliography of that literature. After going largely quiescent for nearly 20 years, interest in the topic resurfaced again starting roughly in the mid-2000s, coinciding with the rise of interest in complex networks generally. Nevertheless, much work of a statistical nature remains to be done in this area.

Formally, we will focus on the following context. Assume there is a system under study that may be represented by a network graph G, which we will call the *population graph*. Suppose in addition, however, that instead of having all of G available to us, we take measurements that effectively yield a sample of vertices and edges, which we compile into a graph $G^* = (V^*, E^*)$. We will refer to G^* as a *sampled graph*. The sampled graph G^* will often be a subgraph of G (i.e., $V^* \subseteq V$ and $E^* \subseteq E$), although this will not always be the case. For example, in situations where there is error in assessing the existence of vertices or edges, there is a chance of spurious observations.[1]

Now suppose that there is a particular characteristic of G, denoted $\eta(G)$, that is of interest. For example, $\eta(G)$ might be a structural characteristic of G, such as the number of edges N_e, the average degree, or the distribution of vertex betweenness centrality scores. Similarly, it might be a summary of some quantity with which the nodes or edges in G are decorated, such as the proportion of men with more female than male friends in a social network. If G is sampled, then typically it will be impossible to recover the exact value of $\eta(G)$ from only the partial information supplied by G^*. The question thus arises as to whether we may still obtain a useful estimate of $\eta(G)$, say $\hat{\eta}$, from G^*.

Intuitively, it is attractive to think that we might simply estimate $\eta(G)$ by $\hat{\eta} = \eta(G^*)$. That is, we might use a 'plug-in' method and estimate the characteristic of interest by the value of that characteristic observed in the sampled graph G^*. This approach is in fact implicitly what is used in any network study that asserts that properties of an observed network graph are representative of those same properties for the graph of the network from which the data were sampled. And certainly many familiar estimators in general practice are plug-in estimators of this type. Sample means, standard deviations, and quantiles, for example, are all both natural and valid estimates of their population

[1] See Section 5.3 for discussion of work in this relatively nascent topic area of 'noisy networks.'

equivalents under standard assumptions of a sample with independent and identically distributed observations.

Unfortunately, in estimating graph characteristics from sampled graphs, this line of reasoning can go awry, thus rendering the reliability of plug-in estimates a nontrivial issue in general. Essentially, at the heart of the trouble is the fact that network sampling often yields unequal probability sampling in the units of interest. This issue is compounded by the fact that, in a typical network analysis study, the summaries of network structures examined may run through multiple types of units (i.e., vertices as units, edges as units, paths as units, etc.). In seeking to address the challenges in this area, recent efforts in the literature have focused to date primarily on the design of sampling strategies intended to reduce the bias of standard estimators resulting from unequal probability sampling. Comparatively less effort has gone towards developing improved estimators for a given sampling strategy. Accordingly, we emphasize the latter of these two directions in this chapter.

While there are many different network structural properties which it may be of interest to estimate, here in this chapter we will focus exclusively on problems involving vertex degrees. Specifically, following some necessary background in Section 3.2, we will look at three successive problems of increasing difficulty: estimation of the mean degree, of the degree distribution, and of the individual vertex degree. The intent behind this approach is to provide insight into the nature of the estimation problem we face in network sampling within just one class of summaries of network structure, in a way that illustrates the increasing difficulty as the scale of the structure of interest goes from global to local.

Some discussion of other important topics in network sampling may be found at the end of this chapter, in Section 3.4. These include the impact of sampling on network modeling and the use of networks to facilitate choice of units in more traditional sampling and estimation problems (e.g., in estimating the proportion of a hard-to-reach population with, say, a given disease).

3.2 Background

3.2.1 An Illustration

We begin with an illustration of some of the ideas just discussed. Suppose that the characteristic of interest is the average degree of a graph G,

$$\eta(G) = (1/N_v) \sum_{i \in V} d_i. \tag{3.1}$$

Let our sample graph G^* be based on the n vertices $V^* = \{i_1, \ldots, i_n\}$, and denote its observed degree sequence by $\{d_i^*\}_{i \in V^*}$. The plug-in estimator of $\eta(G)$ in (3.1) is just the average of the observed degree sequence,

$$\hat{\eta} = \eta(G^*) = (1/n) \sum_{i \in V^*} d_i^*. \tag{3.2}$$

To evaluate this estimator, we consider two sampling designs by which G^* might be obtained. In both cases, we begin with a simple random sample without replacement of n vertices $V^* = \{i_1, \ldots, i_n\}$. Then, in Design 1, for each vertex $i \in V^*$, we observe all edges $\{i,j\} \in E$ involving i; each such edge becomes an element of E^*. On the other hand, in Design 2, we only observe, for each pair $i,j \in V^*$, whether or not $\{i,j\} \in E$; if it is, that edge becomes an element of E^*. Hence, both designs consist of two steps – first sampling a set of vertices V^* and then observing a set of edges E^* – but differ in the manner in which the edges are observed. In either case, the final sampled graph is simply $G^* = (V^*, E^*)$.

Figure 3.1 shows histograms showing the results of calculating the sample average $\hat{\eta} = \eta(G^*)$ under each of these two sampling designs, in the case where the true graph G is taken to be a network of protein interactions in yeast. Here, G consists of $N_v = 5151$ vertices and $N_e = 31{,}201$ edges, and hence has an average degree of $\eta(G) = 12.115$. A random sample of $n = 1500$ vertices was drawn, and edges were sampled as per the specifications of Designs 1 and 2, respectively. This process was repeated for 10,000 trials.

Looking at the figure, it is easy to see that under Design 1, the plug-in estimator $\hat{\eta}$ is quite accurate, with a mean of 12.117 and a standard error of 0.3797. However, under Design 2, it has a substantial bias, with a mean of only 3.528, although its standard deviation is a bit smaller, at 0.2260. The difference in the performance of the estimators in this example is, of course, a function of the difference in the observed degrees $\{d_i^*\}_{i \in V^*}$, induced by the difference in which edges are observed. In the case of Design 1, we observe the actual degree of a vertex $i \in V^*$ (i.e., $d_i^* = d_i$). But in the case of Design 2, we observe a degree that typically under-shoots the actual degree, by a factor of

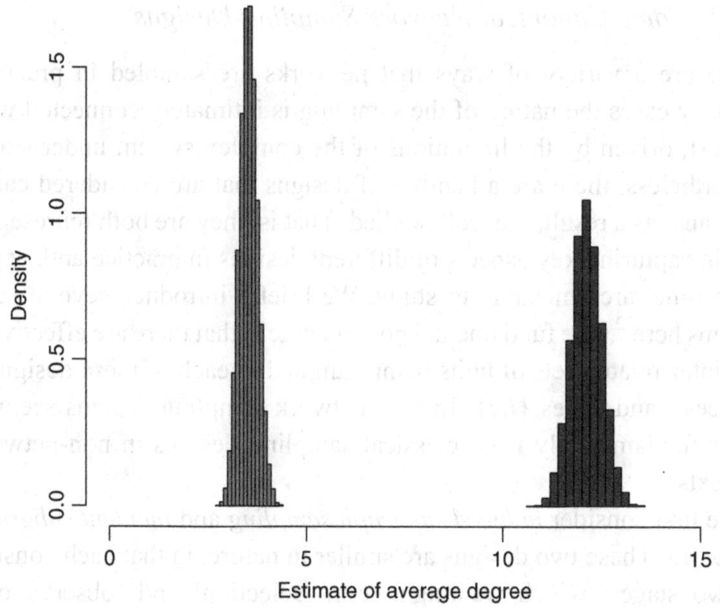

Figure 3.1 Histograms of estimated average degree in a yeast protein interaction network, based on sampling under Design 1 (blue) and Design 2 (red), over 10,000 trials.

roughly n/N_v (i.e., $d_i^* \approx nd_i/N_v$). Hence, despite the fact that in both cases vertices are drawn through simple random sampling, in the latter design our estimate of the average degree is an under-estimate.

While the sampling designs in this illustration are relatively simple, they are by no means uncommon, and rather than being an exceptional case, the behavior witnessed above has been seen to occur under a variety of sampling designs for various choices of characteristic $\eta(G)$. When viewed from the perspective of classical statistical sampling theory, such problems with network graph estimation come as little surprise and, indeed, can be compensated for in principle at least through appropriately designed estimation strategies. In the above illustration, for instance, adjusting the plug-in estimator $\hat{\eta}$ upward by a factor of $N_v/n = 5151/1500 \approx 3.434$ yields an estimator with mean 12.115. In this case the correction is easy, but in general, deriving such correction strategies can be more challenging, depending on the manner in which the topology of the graph G, the characteristics of $\eta(\cdot)$, and the nature of the sampling design interact.

3.2.2 Some Canonical Network Sampling Designs

There are a variety of ways that networks are sampled in practice. In many cases the nature of the sampling is intimately connected with context, driven by the limitations of the complex system under study. Nevertheless, there are a handful of designs that are considered canonical and, as a result, are well-studied. That is, they are both representative in capturing key aspects of different designs in practice and, at the same time, are amenable to study. We briefly introduce several such designs here.[2] One fundamental point to note is that there are effectively two inter-related sets of units being sampled in each of these designs – vertices i and edges $\{i,j\}$. In this, network sampling designs seem to differ fundamentally from classical sampling designs in non-network contexts.

We first consider *induced subgraph sampling* and *incident subgraph sampling*. These two designs are similar in nature, in that each consists of two stages, which we might term ‘selection’ and ‘observation.’ Selection is made among one class of units (i.e., either vertices or edges), which then leads to observation of units from the other class (i.e., edges or vertices). These two complementary designs are shown schematically in Figure 3.2.

Induced subgraph sampling consists of taking a random sample of vertices in a graph G and observing their induced subgraph. More precisely, in this design a simple random sample of n vertices is selected from V, without replacement, which yields the set $V^* = \{i_1, \ldots, i_n\}$. Edges are then observed for all vertex pairs $i,j \in V^*$ for which $\{i,j\} \in E$, yielding the set E^*. Note that this same design was encountered already in the illustration above (i.e., as Design 2). This type of sampling is representative of, say, the construction of contact networks in social network research, when a sample of individuals is first selected and then the individuals are interviewed regarding some measure of contact among themselves (e.g., friendship, likes or dislikes, etc.).

Conversely, under incident subgraph sampling, instead of selecting n vertices in the initial stage, n edges are selected, again through simple random sampling without replacement, directly yielding the set E^*. All vertices incident to the selected edges are then considered observed as

[2] For convenience, our treatment is entirely in terms of undirected graphs, but most of the designs can be generalized in a straightforward fashion to the case of directed graphs.

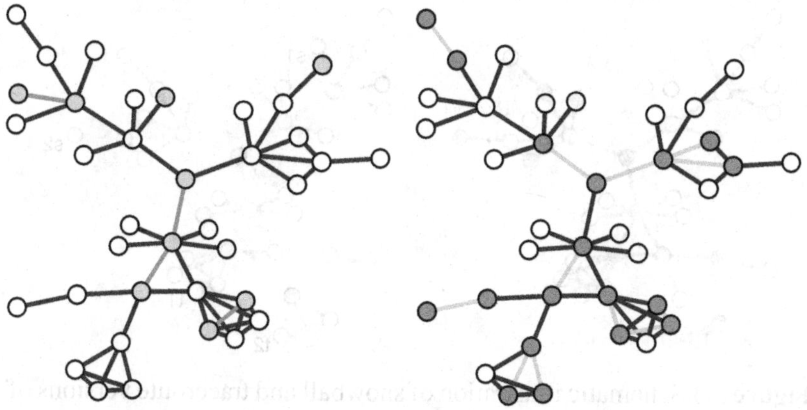

Figure 3.2 Schematic illustration of induced subgraph sampling and incident subgraph sampling. *Left:* Under induced subgraph sampling, a subset of vertices is selected at random (yellow), and edges incident to pairs of nodes in the sample are observed (orange). *Right:* Under incident subgraph sampling, a subset of edges is selected at random (yellow), and vertices incident to these edges are observed (orange).

well, thus providing V^*. Such a design is, for example, implicit in the construction of sampled telephone call graphs, wherein telephone calls are sampled from a database, after which the phone numbers of the initiator and the receiver of the call are observed.

Alternatively, there are various network sampling designs that fall under the general label of *link tracing*. Under such designs, after the selection of an initial sample (typically of vertices), some subset of the edges ('links') from vertices in this sample are traced to additional vertices. Note that, unlike the induced and incident subgraph sampling designs, link-tracing designs generally are iterative, in the sense that they involve multiple stages of sampling, each successive stage of which is dependent upon the previous stage. Two important examples of link-tracing designs are *snowball sampling* and *traceroute sampling*. These are shown schematically in Figure 3.3.

Under snowball sampling, an initial vertex sample, say V_0^*, is extended to its neighboring vertices along the incident edges in between. The process is then repeated some number of times. Formally, define $\mathcal{N}(S)$ to be the set of all neighbors of vertices in a set S. Then snowball sampling extends V_0^* to $V_1^* = \mathcal{N}(V_0^*) \cap \overline{V}_0^*$, and extends V_1^* to $V_2^* = \mathcal{N}(V_1^*) \cap \overline{V}_0^* \cap \overline{V}_1^*$, and so on. The set V_k^* is

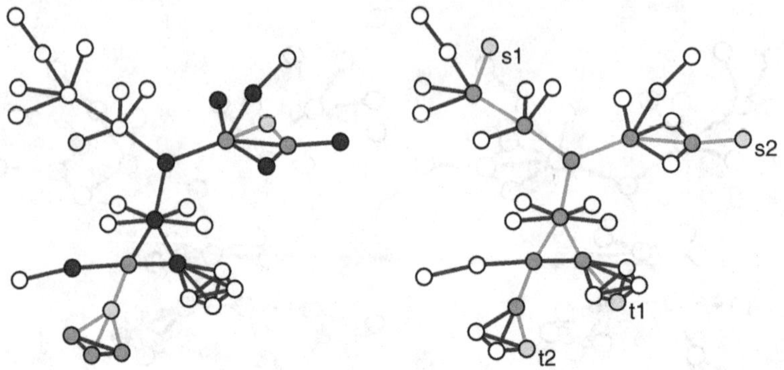

Figure 3.3 Schematic illustration of snowball and traceroute versions of link-tracing designs. *Left:* Under two-stage snowball sampling, vertices are initially selected at random (yellow), after which edges and vertices are observed successively in the first and second waves of sampling outwards (orange and dark red, respectively). *Right:* Under traceroute sampling, paths are traced from selected source vertices $\{s_1, s_2\}$ to selected target vertices $\{t_1, t_2\}$ (yellow), while vertices and edges between are observed (orange).

called the kth *wave* of the sampling process, and $\overline{V}_k^*$ denotes its complement. Sampling can be continued until a wave V_k^* is reached that is empty, or it can be stopped after some number K stages. The final graph G^* obtained through snowball sampling consists of the vertices in $V^* = V_0^* \cup V_1^* \cup \ldots \cup V_K^*$ and, by construction, their incident edges.

Snowball sampling traditionally has been used in social network analysis, where a person is asked to name friends, who are then asked to name friends, and so on and so forth. Some surveys of the World Wide Web graph can be thought of as arising through a variant of snowball sampling. Computer programs called 'spiders' are written to follow an initially compiled subset V_0^* of web pages to those pages corresponding to the HTML addresses listed in the V_0^* pages. The newly discovered pages constitute V_1^*, and these in turn are then examined for new HTML addresses, which are then pursued. In other words, the 'spiders' mimic a human using a web browser by exhaustively following hyperlinks on discovered web pages.[3]

[3] Hyperlinks are directional, with one web page referencing another, and thus the web graph is more properly considered as a directed graph, a detail we have ignored here for the sake of exposition.

An idealized version of the traceroute sampling design is as follows. A sample $S = \{s_1, \ldots, s_{n_s}\}$ of n_s 'sources' is selected from the vertex set V of a network graph G. Then, a sample $T = \{t_1, \ldots, t_{n_t}\}$ of n_t 'targets' is selected from $V \backslash S$. Finally, a route is traced from each source node in S to each target node in T. That is, effectively a path is sampled between each pair $(s_i, t_j) \in S \times T$, and all vertices and edges in the paths are observed. The sampled graph $G^* = (V^*, E^*)$ is then constructed as the union of vertices and edges over all sampled paths.

The term 'traceroute' comes from the name for a computer protocol (called traceroute) that traditionally underlies Internet topology surveys. A similar notion of sampling underlies the design of the 'small-world' studies of Harvard sociologist Stanley Milgram [153], which led later to the play of Guare [97], based on the suggestion that we are each separated from any other person on the planet by at most six other people (i.e., 'six degrees'). Watts and colleagues [64] later conducted a modern version of Milgram's experiment in the context of the Internet.

3.2.3 Interaction of Design, Network, and Summary

Part of what appears to make the problem of inference under network sampling challenging is the apparent interaction of (i) network sampling design, (ii) network topology, and (iii) choice of network summary characteristic. This interaction has been remarked upon in the literature, based on empirical evidence, but its detailed characterization over the 'space' spanned by choice of design, network, and summary is currently lacking.

An early empirical study on this topic was done by Lee and colleagues [133]. The focus of the study was on assessing the bias of various plug-in estimators. They simulated three types of sampling (induced subgraph, incident subgraph, and snowball sampling), on four types of network (Barabasi–Albert random graph, protein–protein interaction network in yeast, an Internet network at the level of autonomous systems, and an arXiv coauthorship network), and evaluated the accuracy of five different numerical summaries of network topology. The results of their study are summarized in Table 3.1. Generally speaking, while arguably there are certain subpatterns evident in the table, it seems difficult to conclude that there is any single global pattern present.

Table 3.1 Visual summary of results from [133]. Entries indicate direction of estimation bias for induced subgraph (red), incident subgraph (green), and snowball (blue) sampling, as a function of choice of network (columns) and network summary (rows).

	BA	**PPI**	**AS**	**arXiv**
Degree Exponent	$\uparrow\uparrow\downarrow$	$\uparrow\uparrow=$	$==\downarrow$	$\uparrow\uparrow\downarrow$
Average Path Length	$\uparrow\uparrow=$	$\uparrow\uparrow\downarrow$	$\uparrow\uparrow\downarrow$	$\uparrow\uparrow\downarrow$
Betweenness	$\uparrow\uparrow\downarrow$	$\uparrow\uparrow\downarrow$	$\uparrow\uparrow\downarrow$	$===$
Assortativity	$==\downarrow$	$==\downarrow$	$==\downarrow$	$==\downarrow$
Clustering Coefficient	$==\uparrow$	$\uparrow\downarrow\uparrow$	$\downarrow\downarrow\uparrow$	$\downarrow\downarrow\downarrow$

As a result, progress in this area presumably will need to come from better understanding of key subproblems (i.e., based on specific sampling plans, particular classes of network summaries, etc.). From a theoretical perspective, some important initial progress towards this goal has been made in recent years for the case of induced subgraph sampling. Work of this sort has been closely intertwined with work on graph limits and, in turn, is motivated by and has close connections to another related literature (i.e., that on so-called property testing in graphs). See [167], for example, for a survey on this latter topic.

Seminal results of this nature may be found in the work of Borgs and colleagues [36]. There the authors define the notion of what they call a 'testable graph parameter.' Let G_n^* be a subgraph of a graph G obtained through induced subgraph sampling, based on a sample of n vertices selected uniformly without replacement. In [36], a network summary statistic η is said to be 'testable' if for every $\varepsilon > 0$ there is a sample size n such that for any graph G with $N_v \geq n$, an estimate $\hat{\eta} = \hat{\eta}(G^*)$ can be computed such that

$$\mathbb{P}(|\eta(G) - \hat{\eta}| > \varepsilon) \leq \varepsilon. \tag{3.3}$$

A natural candidate is the plug-in estimator $\eta\left(G_n^*\right)$. In other words, testability of a summary η is essentially equivalent to the existence of an estimator of η satisfying a weak form of consistency. Hence, an understanding of what summaries η are testable can provide insight into those cases wherein we may expect that simple estimates of η under induced subgraph sampling will be accurate.

Importantly, in [36] there are several equivalent criteria given for establishing testability of a specific choice of summary η. Key to these various criteria is the result [36, Prop 2.12(a)] that η is testable if and only if $\eta(G_n)$ converges for every convergent graph sequence $G_1, G_2, \ldots$. Convergence of graph sequences is defined with respect to the cut-distance $d_\square(\cdot,\cdot)$, as defined in (2.6). And accuracy of approximation through induced subgraph sampling is characterized in this distance through the following result [36, Thm 2.9]:

$$d_\square(G, G_n^*) \le \frac{10}{\sqrt{\log_2 n}} \tag{3.4}$$

with probability at least $1 - e^{-n^2/(2\log_2 n)}$. In words, under induced subgraph sampling, the sampled graph G_n^* will be close to G with high probability for sufficiently large n. Thus, if induced subgraph sampling under sufficiently large n produces sufficiently accurate approximations to the underlying graph G, then, intuitively, we can hope that accuracy of estimation is feasible for sufficiently 'smooth' summaries η.

Combined, the above results suggest a possible path forward for approaching estimation problems under induced subgraph sampling – particularly for determining feasibility of a given problem. To the best of our knowledge, this path has yet to be explored in depth for the practical development of specific estimators and specific classes of network summaries η. More generally, it would be of interest to see to what extent a similar path can be established for other network sampling designs, with incident subgraph sampling being the likely next candidate.

3.3 Case Study: Estimating Degree-Based Characteristics

While admittedly only preliminary, the story emerging from the discussion of the previous section is encouraging, in that it suggests there is hope for producing a general and comprehensive theory characterizing the aforementioned interaction of network sampling, network topology, and network summary. Until that point is reached, however, much insight presumably is to be obtained through the study of individual classes of network sampling and inference problems. Here in this section we will focus on one specific type of network characteristic – vertex degree. And, for the purposes of illustration, we will focus on one particular network sampling design – induced subgraph sampling.

We will consider the estimation problem at three scales at which degree information in a network may be desired: through the average vertex degree, through the degree distribution, and through the degree sequence (i.e., individual vertex degrees).

3.3.1 Average Vertex Degree

The problem of estimating the average vertex degree (3.1) under induced subgraph sampling is the same estimation problem encountered already in the illustration in Section 3.2.1, under the so-called Design 2. There we noted in hindsight that a correction of the simple average degree (3.2) in the sampled network G^*, by a factor n/N_v, appeared to be sufficient for correcting the bias of this plug-in estimator. Such an approach has been developed more formally by Frank [76], which we describe now.

First, note that it is sufficient to consider the problem of estimating the total number of edges N_e in the network graph G, since the average degree $\eta(G)$ is equal to $2N_e/N_v$. (This presumes, of course, that the number of vertices N_v is known.) Frank derives explicit formulas for the Horvitz–Thompson estimator of N_e, its variance, and an unbiased estimate of this variance.[4] Central to these formulas are vertex inclusion probabilities up to fourth order, which, fortuitously, depend only on the number of distinct vertices in certain subsets (i.e., specifically, within vertex subsets of order one, two, three, or four) and not on the actual vertices themselves. In particular, these formulas rely on the values

$$p_r = \binom{n}{r} \Big/ \binom{N_v}{r}, \tag{3.5}$$

where $p_r = \mathbb{P}(\{i_1, \ldots, i_r\} \in S)$, for $S = \{i_1, \ldots, i_n\}$ the subset of vertices sampled under the induced subgraph sampling design and $r = 1, 2, 3$, or 4.

[4] Horvitz–Thompson estimation is a strategy from classical sampling theory that allows for the unbiased estimation of totals and averages under generic (i.e., potentially unequal probability) sampling designs through the use of weighted averages. Key to this approach is the availability of the so-called inclusion probabilities to use as weights – that is, the probabilities π_i that a given unit i in the underlying population is included in the sample under a given sampling design (as well as higher-order variants thereof). See [195], for example, for background.

The estimator of N_e that emerges from this approach is just

$$\hat{N}_e = p_2^{-1} N_e^*, \tag{3.6}$$

where N_e^* is the number of edges in the sampled subgraph G^*. This estimator simply scales up the empirically observed total N_e^* by a factor p_2^{-1}. The variance of this estimator takes the form

$$\mathbb{V}(\hat{N}_e) = \alpha_0 N_e^2 + \alpha_1 Q + \alpha_2 N_e, \tag{3.7}$$

where $Q = \sum_{i \in V} d_i^2$ is the sum of squares of the vertex degrees in G and

$$\begin{aligned} \alpha_0 &= (p_4 - p_2^2)/p_2^2, \\ \alpha_1 &= (p_3 - p_4)/p_2^2, \\ \alpha_2 &= (p_2 - 2p_3 + p_4)/p_2^2. \end{aligned} \tag{3.8}$$

An unbiased estimate of this variance is given by

$$\hat{\mathbb{V}}(\hat{N}_e) = \beta_0 N_e^{*^2} + \beta_1 Q^* + \beta_2 N_e^*, \tag{3.9}$$

where Q^* is the analogue of Q on G^* and

$$\begin{aligned} \beta_0 &= \frac{1}{p_2^2} - \frac{1}{p_4}, \\ \beta_1 &= \frac{1}{p_4} - \frac{1}{p_3}, \\ \beta_2 &= \frac{2}{p_3} - \frac{1}{p_2} - \frac{1}{p_4}. \end{aligned} \tag{3.10}$$

Figure 3.4 shows the results of a numerical simulation, in which the true graph G is the same network of protein interactions as in the illustration in Section 3.2.1. Induced subgraph sampling was simulated in each of 10,000 trials, using Bernoulli sampling of vertices with $p = 0.10, 0.20$, and 0.30. Shown in the figure are histograms of the estimators $\hat{N}_e$ in (3.6), for each choice of p, and of the estimated standard errors of these estimators, based on (3.9). The average of $\hat{N}_e$ over the simulations was 31116, 31197, and 31203, for $p = 0.10, 0.20$, and 0.30, respectively. Thus, the unbiasedness of all three estimators in estimating $N_e = 31,201$ is well-supported by these results. The unbiasedness of the estimated variances (3.9) was similarly supported.

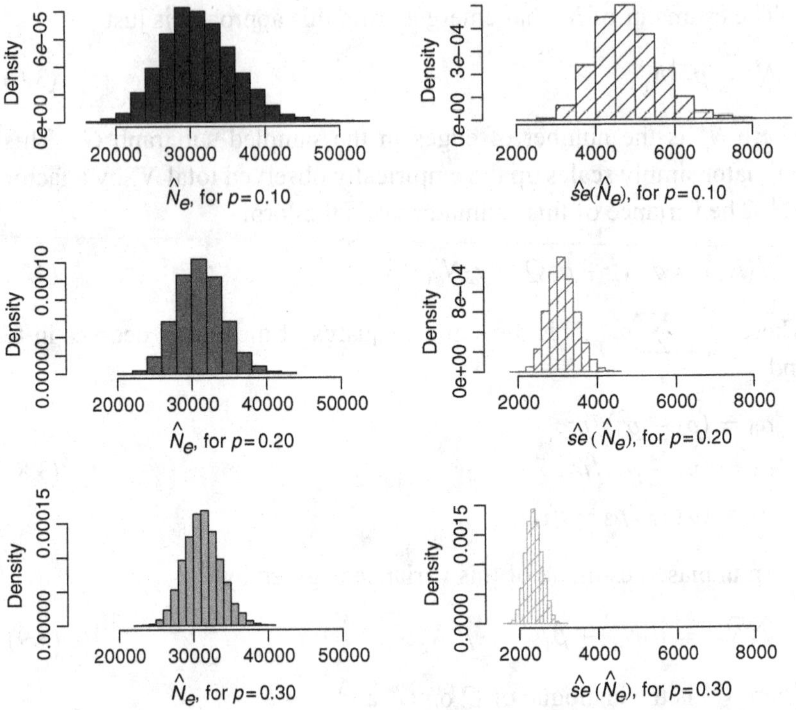

Figure 3.4 Histograms of estimates $\hat{N}_e$ (left) of $N_e = 31,201$, as well as estimated standard errors (right), in a yeast protein interaction network, under induced subgraph sampling, with Bernoulli sampling of vertices, using $p = 0.10$ (blue), 0.20 (red), and 0.30 (yellow). Results based on 10,000 trials.

Frank [78] offers results similar to those above for the estimation of certain other totals in graphs. See [127, Ch 5.4] for an overview of the use of Horvitz–Thompson principles in estimating graph totals. At present there does not appear to be a similarly comprehensive treatment for network summaries η that fall outside the Horvitz–Thompson framework (i.e., do not take the form of totals or means).

3.3.2 Network Degree Distribution

The average degree of a network is a fairly crude summary of overall connectivity. Of more standard interest in the literature is the degree distribution. The degree distribution of a network G, denoted by $\{f_d\}$, specifies the proportion f_d of vertices having exactly d incident edges, for $d = 0, 1, \ldots$. It is considered a fundamental quantity associated with a network. In particular, the shape (especially the tail behavior) can have profound implications on a variety of other characteristics and properties of the network, including the behavior of network-indexed processes like disease or rumor spread. And, importantly, it is known that the degree distribution may be adversely affected by sampling, sometimes dramatically so (e.g., [132, 187, 188]). Hence, the task of recovering the degree distribution of some true underlying network G, given only the information provided by the sampled network G^*, is a statistical task of central importance.

Relatively little work appears to have been done on this problem to date. The first such work appears to have been published in the early 1980s by Ove Frank [79, 80], and it is on this perspective that we focus here. Frank shows that, under certain network sampling designs, the expectation of the observed degree frequencies is a linear combination of the true degree frequencies. Specifically, let $\mathbf{f} = (f_d)$ be the vector of true degree frequencies in G, and $\mathbf{f}^* = (f_d^*)$ the (random) vector of observed degree frequencies in G^*. Then, under a number of standard network sampling designs,

$$\mathbb{E}[\mathbf{f}^*] = \widetilde{P}\mathbf{f}, \tag{3.11}$$

where $\widetilde{P}$ depends fully on the sampling scheme and not on the network itself. Frank observed that a natural unbiased estimator of $\mathbf{f}$ thus would seem to be simply $\widetilde{P}^{-1}\mathbf{f}^*$. However, this estimator was found to suffer from two issues: (i) $\widetilde{P}$ typically is poorly conditioned in practice, and (ii) $\widetilde{P}^{-1}\mathbf{f}^*$ may not be non-negative.

Recently – almost 35 years later – Zhang and colleagues [212] observed that from the perspective of modern nonparametric density estimation, what we face in (3.11) is a linear inverse problem. One which may potentially be quite ill-posed, in the sense that the matrix $\widetilde{P}$ can be quite ill-conditioned. Their solution was to propose a constrained, penalized weighted-least-squares estimator, which, in particular, produces estimates that are non-negative (by constraint) and inverts the matrix $\widetilde{P}$ in a stable fashion (by construction), in a manner that encourages smooth solutions (through a penalty). Below we describe this estimator, for the specific case of induced subgraph sampling.

Without loss of generality, we take as our goal accurate estimation of the vector of degree counts $\mathbf{n} = (n_0, n_1, \ldots, n_M)$, where $n_k = N_v f_k$ is the number of vertices of degree k and M is the maximum degree in the true network G. In principle, the largest possible value for M is $N_v - 1$ in a simple network where no multiple edges or self-loops exist, although in practice we may have knowledge that it is smaller. The equation

$$\mathbb{E}[\mathbf{n}^*] = P\mathbf{n} \tag{3.12}$$

holds in some generality, in analogy to (3.11), where $\mathbf{n}^* = (n_0^*, n_1^*, \ldots, n_M^*)$ is the vector of observed degree counts in G^* and P replaces $\widetilde{P}$.

The (i,j) th entry P_{ij} of the matrix P is the probability that a vertex of degree j in G is selected through random sampling and observed to have degree i in G^*. Suppose that the vertices in our induced subgraph are selected through Bernoulli random sampling with rate p (i.e., rather than uniformly at random without replacement, as earlier in this chapter). Then, the entries P_{ij} depend only on the sampling design (and not on any unknown structure of G), taking the form

$$P_{ij} = \begin{cases} \binom{j}{i} p^{i+1}(1-p)^{j-i} & \text{for } 0 \leq i \leq j \leq M, \\ 0 & \text{for } 0 \leq j < i \leq M. \end{cases} \tag{3.13}$$

Note that P is an upper-triangular matrix. The condition number of this matrix is equal to p^M and so, as the sampling rate p goes down or the maximum degree M increases, the operator P becomes more ill-conditioned. In real-world situations, such as the monitoring of online social networks, sampling rates are typically low (e.g., 10–20%) and M is typically large (e.g., on the order of 100s or 1000s), and thus P is decidedly ill-conditioned and effectively not invertible.

The naive estimator $P^{-1}\mathbf{n}^*$ of $\mathbf{n}$ motivated by (3.12) can be viewed as a least-squares solution, under the model $\mathbf{n}^* = P\mathbf{n} + \mathbf{e}$, where $\mathbf{e} = \mathbf{n}^* - \mathbb{E}[\mathbf{n}^*]$. In fact, however, the covariance $C = \mathrm{Cov}(\mathbf{e})$ can be shown to have nonconstant variance and nonzero off-diagonal entries. So, a generalized least-squares approach would seem preferable. And, given the ill-conditioned nature of P, some regularization is required. Accordingly, Zhang and colleagues propose the estimator $\hat{\mathbf{n}}$ defined as the solution to the following optimization problem:

$$\begin{aligned} \underset{\mathbf{n}}{\text{minimize}} \quad & (P\mathbf{n} - \mathbf{n}^*)^T C^{-1} (P\mathbf{n} - \mathbf{n}^*) + \lambda \cdot \text{pen}(\mathbf{n}) \\ \text{subject to} \quad & n_i \geq 0,\ i = 0, 1, \ldots, M \\ & \sum_{i=0}^{M} n_i = n_v, \end{aligned} \tag{3.14}$$

where C is as above, pen($\mathbf{n}$) is a penalty on the complexity of a candidate value $\mathbf{n}$, and λ is a smoothing parameter.

Under a convex choice of penalty, (3.14) has the canonical form of a convex optimization and, in principle, standard software can be used for its solution. The assumption of a smooth true degree distribution is natural, and can be accounted for by choosing a penalization of the form $\| D\mathbf{n} \|_2^2$, where the matrix D represents a second-order differencing operator. Specifically, the formula for D is

$$D = \begin{bmatrix} 1 & -2 & 1 & 0 & \ldots & 0 & 0 & 0 & 0 \\ 0 & 1 & -2 & 1 & \ldots & 0 & 0 & 0 & 0 \\ 0 & 0 & 1 & -2 & \ldots & 0 & 0 & 0 & 0 \\ \vdots & \vdots & \vdots & \vdots & & \vdots & \vdots & \vdots & \vdots \\ 0 & 0 & 0 & 0 & \ldots & -2 & 1 & 0 & 0 \\ 0 & 0 & 0 & 0 & \ldots & 1 & -2 & 1 & 0 \\ 0 & 0 & 0 & 0 & \ldots & 0 & 1 & -2 & 1 \end{bmatrix}. \tag{3.15}$$

This choice, here in the discrete setting, is analogous to the use of a Sobolev norm with nonparametric density estimation in the continuous setting. The resulting optimization defining $\hat{\mathbf{n}}$ can be written as a quadratic programming problem.

The problem of selecting an optimal λ turns out to be nontrivial in the present context. Commonly used cross-validation methods which assume independent and identically distributed observations do not apply to our network sampling situation, since the elements of the vector $\mathbf{e}$ are neither independent nor identically distributed. But a strategy based on the method of generalized Stein's unbiased risk estimation (SURE), as proposed in [69], may be usefully brought to bear. Specifically, defining a weighted mean square error (WMSE) in the observation space as

$$WMSE(\hat{\mathbf{n}}, \mathbf{n}) = \mathbb{E}\left[(P\mathbf{n} - P\hat{\mathbf{n}})^T C^{-1} (P\mathbf{n} - P\hat{\mathbf{n}})\right], \tag{3.16}$$

we aim to select that λ which minimizes the WMSE in $\hat{\mathbf{n}} \equiv \hat{\mathbf{n}}(\lambda)$. While the WMSE itself is unknown to us, under appropriate conditions a generalized SURE estimate can be obtained as

$$\widehat{WMSE}(\hat{\mathbf{n}}, \mathbf{n}) = (P\mathbf{n})^T C^{-1} P\mathbf{n} + (P\hat{\mathbf{n}})^T C^{-1} P\hat{\mathbf{n}} \\ +2\left\{\text{Trace}\left(P\frac{\partial \hat{\mathbf{n}}}{\partial \mathbf{n}^*}\right)\right\} \\ -2(P\hat{\mathbf{n}})^T C^{-1}\mathbf{n}^*. \tag{3.17}$$

The first term in (3.17) does not involve λ. The last three terms have $\hat{\mathbf{n}}$ in them, which is a function of λ. Given P, $\mathbf{n}^*$, and C as well, the second and fourth terms are straightforward to compute. The third term, called the divergence term in [69], can be simulated using the Monte Carlo technique proposed in [162]. Zhang and colleagues recommend approximating C by a diagonal matrix, based on a simple smoothing of $\mathbf{n}^*$.

An illustration of the performance of the estimator $\hat{\mathbf{n}}$ is shown in Figure 3.5, on data from three online social networks: Friendster, Orkut, and LiveJournal.[5] In these online social networks, users create functional groups that others can join, based on, for example, topics, shared interests and hobbies, or geographical regions. In this illustration, ground-truth communities [207] were used to define three subnetworks for each social media platform, based on a union of the largest five communities, the next 6–15 communities, and the next 16–30 communities, respectively. Although artificial, this device allows for the definition of several networks with different characteristics from each social media platform. The goal was to estimate the degree distribution underlying each subnetwork, for each platform, under induced subgraph sampling. Examination of these plots shows that, while the sampled degree distribution can be quite off from the truth, particularly in the case of the Friendster and Orkut networks, correction for sampling using the methodology described above results in estimates that are nearly indistinguishable by eye from the true degree distributions on the log–log scale at which they are visualized.

Hence, for certain sampling designs, it appears that the estimation of the degree distribution in a nonparametric fashion is not only feasible but can be done with reasonable accuracy. However, a formal characterization of just what level of accuracy to expect,

[5] These data are available on the SNAP (Stanford Network Analysis Project) website.

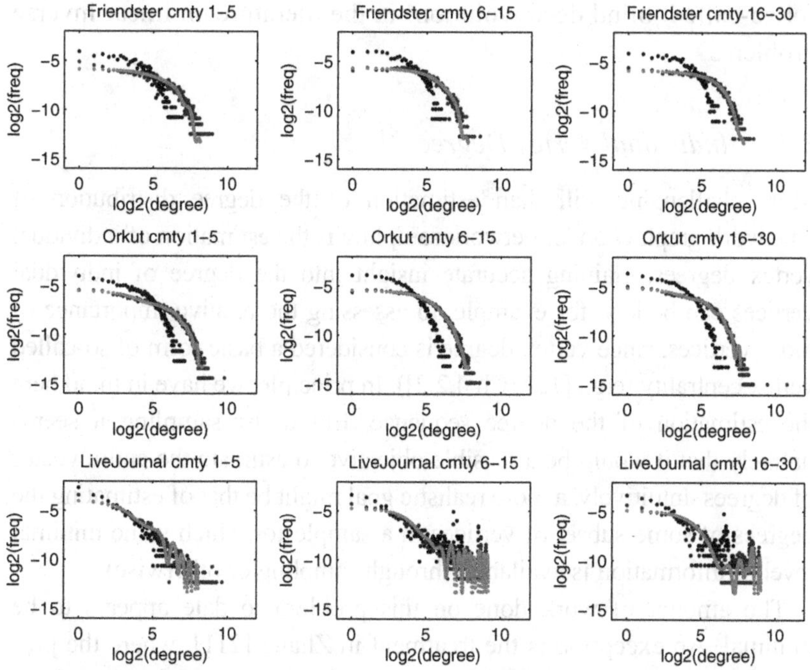

Figure 3.5 Estimating degree distributions of communities from Friendster (top), Orkut (middle), and LiveJournal (bottom), for subnetworks formed from the union of larger (left), medium (center), and smaller (right) ground-truth communities. Blue dots represent the true degree distributions, black dots represent the sample degree distributions, and red dots represent the estimated degree distributions. Sampling rate is 30%. (Dots which correspond to a density $< 10^{-4}$ are eliminated from the plot.)

and under what conditions (e.g., on some combination of sampling rate p, shape of degree distribution $\{f_d\}$, etc.), remains to be determined. Additionally, while the matrix P (respectively, $\widetilde{P}$) in the fundamental relationship in (3.12) (respectively, (3.11)) is known to have a functional form similarly simple to that of induced subgraph sampling for a handful of other common designs (e.g., incident subgraph sampling, one-wave snowball sampling, and certain types of random walk sampling), such cannot be expected to be the case for all designs. For example, in snowball sampling involving more than one wave, even individual vertex-inclusion probabilities become considerably more involved. This might suggest the relevance of methods for so-called 'blind deconvolution' in the literature on linear inverse problems.

3.3.3 Individual Vertex Degree

More challenging still than estimation of the degree distribution of a network graph G under network sampling is the estimation of individual vertex degrees. Gaining accurate insight into the degree of individual vertices can be key, for example, to assessing the relative importance of those vertices, since vertex degree is considered a basic form of so-called vertex centrality (e.g., [127, Ch 4.2.2]). In principle, we have in mind here the estimation of the degree sequence. But under sampling it seems unlikely that it would be a feasible objective to estimate the entire vector of degrees. Intuitively, a more realistic goal might be that of estimating the degrees of some subset of vertices in a sample for which some minimal level of information is available (through sampling or otherwise).

The amount of work done on this problem to date appears to be minimal. An exception is the treatment in Zhang [211], where the problem arises in the form of an illustration of a more general framework for estimating sums of random variables. There the network sampling design considered is specifically traceroute sampling, which yields an estimation problem that can be considered a variant of the so-called species problem [40]. A similarly formal treatment seems not yet to have been developed more generally for other canonical sampling strategies. Arguably most related to the present context is work on the problem of estimating the size of a so-called personal network, say for a group of people in a survey. Work in this direction has, for example, included the proposal of estimators obtained by scaling up the degree observed for vertices in an observed network G^*, although network sampling designs do not seem to play an explicit role in these studies [125, 150].

In order to illustrate the nature of some of the challenges that can be expected to arise in estimating individual vertex degrees, we describe below the results of some preliminary work with Ganguly [86] for the case of induced subgraph sampling. Let $\mathrm{d} = (d_1, \ldots, d_{N_v})^T$ denote the vector of true degrees for the set of vertices $V = \{1, \ldots, N_v\}$ in the underlying network G. Similarly, let $\mathbf{d}^* = (d_1^*, \ldots, d_{N_v}^*)^T$ denote the (random) vector of degrees observed for those same vertices under induced subgraph sampling. Note that, clearly, $d_i^* = 0$ for any vertex $i \notin V^*$. As in Section 3.3.2, suppose that vertices in V^* are obtained through Bernoulli random sampling with rate p. Under this network sampling design, d_i^* is distributed as a binomial random variable, with parameters d_i and p, for each $i = 1, \ldots, N_v$. The first moment is therefore just $\mathbb{E}[\mathbf{d}^*] = p\mathbf{d}$ and a method-of-moments perspective trivially suggests $\hat{\mathbf{d}}^{MM} = p^{-1}\mathbf{d}^*$ as a natural scale-up estimator of $\mathbf{d}$.

It seems likely that some improvement can be made on this estimator, although just how much and in what contexts is not immediately clear. While there are many directions in which one might explore alternative estimators, if information on the degree distribution is available, a Bayesian approach seems natural. Such information might be available either *a priori* through knowledge of the complex system under study or through estimation, as in (3.14), yielding a type of empirical Bayes estimate.

Consider the problem of estimating the degree d_i for a specific vertex i, and suppose that d_i follows a distribution $g(\cdot)$. Then, assuming induced subgraph sampling, the Bayes estimator of d_i under square error loss can be derived easily as

$$\hat{d}_i^B = \frac{\sum_{d_i \geq d_i^*} d_i \binom{d_i}{d_i^*} (1-p)^{d_i} g(d_i)}{\sum_{d_i \geq d_i^*} \binom{d_i}{d_i^*} (1-p)^{d_i} g(d_i)}. \tag{3.18}$$

Now let $\mathcal{R}(\hat{d}_i, d_i)$ denote the (frequentist) risk of an estimator $\hat{d}_i$ of the degree d_i of vertex i. Then it can be shown [86, Prop 3.3] that

$$\mathcal{R}(\hat{d}_i^B, d_i) \leq \mathcal{R}(\hat{d}_i^{MM}, d_i)$$

on the set of sampled graphs G^* for which the following two conditions hold:

$$\mathbb{E}\left(\sum_{d_i' \geq d_i^*} g^2(d_i') \right) \leq \frac{p(1-p)}{(N_v - 1 - d_i)^2} d_i \text{ when } d_i \leq \frac{N_v - 1}{2} \tag{3.19}$$

$$\text{and } \frac{\sum_{d_i \geq d_i^*} p(d_i^*, d_i) g(d_i)}{\sum_{d_i \geq d_i^*} p(d_i^*, d_i)} \geq p, \tag{3.20}$$

where $p\left(d_i^*, d_i\right) = \binom{d_i}{d_i^*}(1-p)^{d_i}$.

The conditions (3.19) and (3.20) essentially constrain the tail behavior of the prior degree distribution. The first condition ensures that the decay is such that the tail is not too 'thick' and the second condition ensures that it is not too 'thin.' Note too that as d_i becomes bigger, the right-hand side in condition (3.19) becomes smaller, meaning that relatively few nodes can have very high degree, a property consistent with sparse graphs. Finally, the left-hand side in the condition (3.20) can be interpreted as the mean of the tail probabilities weighted by the posterior distribution. This has to be bounded away from zero in order for the Bayes estimator to have lower risk than the simple method-of-moments estimator.

Simulations suggest that the improvement offered by the Bayes estimator can range from modest to substantial, as shown in Tables 3.2 and 3.3. For a graph G with a concentrated degree distribution g, as in the case of the Erdös–Rényi graph underlying the results in Table 3.2, the improvement is uniformly decisive, with the risk of the Bayes estimator found to be anywhere from half to nearly a quarter that of the method-of-moments estimator. On the other hand, for a graph with a diffuse degree distribution, as in the case of the scale-free model used to produce the results in Table 3.3, the sparseness of G appears to play an important role. Specifically, while the risk of the Bayes estimator is roughly half that of the method-of-moments estimator for the sparsest case considered, the gap between the two narrows considerably

Table 3.2 Comparison of risk $\mathcal{R}(\hat{d}, d)$ over all vertices for the method-of-moments estimator, $\hat{d}^{MM}$, and the Bayes estimator, $\hat{d}^{B}$, under induced subgraph sampling with $p = 0.10$ or 0.20, when G is an Erdös–Rényi graph with $N_v = 1000$ vertices and density δ.

δ, p	MM	Bayes
$\delta = 0.1, p = 0.1$	292.29	90.03
$\delta = 0.2, p = 0.1$	416.02	121.32
$\delta = 0.3, p = 0.1$	492.22	136.86
$\delta = 0.4, p = 0.1$	588.18	152.94
$\delta = 0.1, p = 0.2$	284.08	119.87
$\delta = 0.2, p = 0.2$	389.15	164.30
$\delta = 0.3, p = 0.2$	485.09	187.43
$\delta = 0.4, p = 0.2$	527.37	205.47

Table 3.3 Comparison of risk $\mathcal{R}(\hat{d}, d)$ over all vertices for the method-of-moments estimator, $\hat{d}^{MM}$, and the Bayes estimator, $\hat{d}^{B}$, under induced subgraph sampling with $p = 0.10$, when G is a scale-free random graph with $N_v = 1000$ vertices, density δ, and tail parameter m (with larger m corresponding to lighter tails).

δ, m	**MM**	**Bayes**
$\delta = 0.002, m = 2$	45.60	33.21
$\delta = 0.01, m = 2$	92.13	82.29
$\delta = 0.05, m = 2$	238.10	232.76
$\delta = 0.002, m = 2.5$	42.48	19.23
$\delta = 0.01, m = 2.5$	92.91	81.93
$\delta = 0.05, m = 2.5$	210.04	231.68
$\delta = 0.002, m = 3$	41.52	21.71
$\delta = 0.01, m = 3$	89.40	83.39
$\delta = 0.05, m = 3$	209.97	242.90

as the graph grows more dense, until finally, method-of-moments outperforms Bayes.

These numerical results appear to be reasonably robust to misspecification of the prior [86]. More formally, characterizations can be given of the accuracy of the corresponding empirical Bayes estimator, for the case where g is replaced in (3.18) with an estimate $\hat{g}$, when the discrepancy with g is no more than some $\varepsilon > 0$.

In general, there is considerable room for work on the problem of estimating the degree sequence under network sampling, including both further development in the case of induced subgraph sampling explored above and analogous work with other sampling designs. The challenge of understanding precisely where improvements can be expected (i.e., whether for all vertices or only some), and under what conditions, seems to be a particularly interesting aspect of this problem.

3.4 Related Topics

Our goal in this chapter was to introduce network sampling in such a way as to provide both a broad sense of the topic at 'ten thousand feet' as it were, and, complementing that, a more in-depth look at a select few topics in one particular context. The hope is that the reader emerges from this chapter not only with an idea of the vast potential for additional work on statistical foundations in this area, but also with some initial feeling for the character of the problems and their inherent challenges.

Indeed, there are a multitude of directions that might be pursued, and certainly there are many threads in the literature along which researchers are doing so already. In the classical literature this includes, for example, work under network sampling on other estimation problems not discussed in this chapter, such as for network density [95], various network totals and subgraph counts (e.g., [76, 78]), and the number of connected components [77]. More recent work in the same spirit includes, for example, the estimation of centrality measures (e.g., [38, 70]). Arguably, however, the preponderance of attention in the modern era has been focused to date on (i) studying the impact (usually in an empirical manner) of sampling on standard network summary statistics $\eta(G^*)$ of interest, and (ii) developing network sampling designs that produce subgraphs that are sufficiently 'close' to the true underlying graph for these statistics (i.e., designs that reduce or eliminate bias in certain respects).

Perhaps the most well-known example of the first of these directions is the work on understanding the implications of sampling on the degree distribution. Specifically, a number of seminal papers in the literature demonstrate the effects that different network sampling designs can have on rendering the observed degree distribution unrepresentative of the true underlying degree distribution, with particular interest in the issue of broad versus concentrated degree distributions. One of the first of these was Lakhina and colleagues [132], who present results from numerical experiments showing that extreme forms of traceroute sampling can actually induce a broad degree distribution in the sampled graph G^* when none exists in the true graph G. Analytical work (e.g., [2, 52]) has confirmed and refined these findings, showing that even when G has an actual power-law degree distribution,[6] the exponent can be significantly under-estimated by exponents obtained from the degree distribution of G^*. Similar results have been demonstrated in the case of sampling protein–protein interaction networks [99]. See also the work of Stumpf and colleagues [187, 188].

Seminal to the second of the directions mentioned above is the work of Leskovec and Faloutsos [135]. The primary contribution there is

[6] That is, when $f_d \propto d^{-\alpha}$.

a formalization of the notion of the extent to which a sampled subgraph G^* is 'representative' of the true underlying graph G. Those authors recommend using the Kolmogorov–Smirnov distance between distributions of various statistics of interest (e.g., the degree distribution, the distribution of sizes of strongly connected components, etc.). Following up on this work, Hübler and colleagues [112] pose the problem as one of optimization and propose a Metropolis-based algorithm for sampling subgraphs in a manner that seeks to minimize a given distance between arbitrary topological properties on G^* and G. Intuitively, however, there must be some natural tension induced between seeking accuracy for certain properties versus others, which can be expected to impact such algorithms, as pointed out by Neville and colleagues [3]. Such tension can be thought of as a manifestation of the same issues discussed here in Section 3.2.3.

In terms of domain areas, there are a handful of specific areas in which there has been a particular interest in network sampling. These are areas in which concerted effort has been made to develop sampling and estimation procedures, typically in a manner closely coupled to particular applications or classes of applications. One such area is networked systems, where interest in monitoring massive engineered systems like peer-to-peer networks, the World Wide Web, and online social networks naturally suggests sampling to balance the need for information about the system against the various costs of obtaining it. Usually, costs in this context reflect some combination of time, resources, and intrusiveness (e.g., revisiting a vertex too frequently under, say, random-walk sampling). Another such area is data mining, where networks are considered a now-canonical data type. It is in this context that much of the work on representative subgraph sampling has been done. Often in this area, success in high-level tasks (e.g., classification problems) is used too as an indication of the effectiveness of sampling. For a short survey of representative publications from the substantial literatures on these first two areas, see [4, Sec 4].

Additionally, a third domain area in which network sampling has received a significant amount of attention is in the context of epidemiological studies surveying so-called hidden populations. These are populations of hard-to-reach individuals, such as the homeless, sex workers, or intravenous drug users. There the focus has been largely on the development and use of respondent-driven sampling (RDS), a variation on link-tracing designs that typically incorporates incentives (e.g., coupons) to encourage members of a given population to recruit additional members. This particular form of sampling yields explorations of the network that, ideally, are conceptualized as random walks, and the corresponding statistical methodology for estimation of population characteristics (e.g., proportion of sex workers with AIDS) tends to draw on principles of Horvitz–Thompson estimation. Originally proposed by Heckathorn [103, 104], with additional key innovations in joint work with Salganick [173] and Volz [200], the RDS methodology has by now been applied in hundreds of studies [119]. However, the biases inherent in the sampling, and the key role played by nontrivial assumptions in producing desirable properties claimed for the resulting estimates, has caused many to voice a note of caution (e.g., [89, 93, 140]). Efforts continue within the statistical community to better understand the behavior of, and to improve upon, RDS (e.g., [88, 90, 137, 141]).

On a final note, we point out the dearth of work on the interplay between the first two topics of this monograph – network modeling and network sampling. While it is generally acknowledged that most empirical network data has an element of sampling to it, the vast majority of network modeling exercises do not explicitly reflect this fact. Rather, as in Chapter 2, the focus in model specification typically is on capturing generative mechanisms for what in this chapter would be termed the 'true' edges (i.e., the elements in E rather than E^*). The absence of substantial attention on this question to date would seem to represent a major gap in the statistical foundations for network science. And the handful of existing works in this area would seem to suggest that the interplay of sampling and modeling for networks can be expected to be both subtle and nontrivial.

For example, building on earlier work of Thompson and Frank [196], Handcock and Gile [101] contrast traditional design-based and model-based perspectives in the context of sampled network data.[7] Offering a network-based analogue of the notion of ignorability in traditional missing-data analysis, they demonstrate that many network sampling designs are ignorable in this sense and show how likelihood-based methods of inference may be used to take advantage

[7] This is a classical distinction in the survey and experimental design literatures. Design-based approaches to inference traditionally view random variation in the sample as coming only from the underlying sampling design and not from, say, uncertainty in measurements or generative mechanisms.

of this fact. See too Lunagomez and Airoldi [142] for recent related work from a Bayesian perspective. Alternatively, Shalizi [176] offers a notion of 'consistency under sampling' of models, drawing on concepts from the theory of stochastic processes, and shows that in the popular class of ERGMs this notion is violated by many popular model formulations.

Ultimately, it seems that a careful and principled integration of the sampling and modeling perspectives is needed, with care regarding both the goals and constraints offered by both. A thoughtful initial step in this direction has been offered recently by Crane and Dempsey [57].

4
Networked Experiments

4.1 Introduction

Across the sciences – social, biological, and physical alike – there is a pervasive interest in evaluating the effect of treatments or interventions of various kinds. Examples of such treatments and effects include the possible impact of a new marketing campaign for a given product, the hope of moderating depression through a new type of psychotherapy, the potential for reducing post-surgical bacterial infections through the introduction of a new antibiotic, or the goal of improving computing speeds through the design of a new microprocessor. In all such settings, the ideal is to evaluate the proposed treatment in a manner unmarred by bias of any sort. On the other hand, nature and circumstances often conspire to make achievement of this ideal difficult (if not impossible) in many cases. As a result, there is by now a vast literature on the design, conduct, and analysis of studies for evaluating the efficacy of treatment.

The prototypical example of such studies arguably is the randomized controlled trial [184], also called A/B testing more recently in the computer science literature [126]. It is by this that we will mean 'experiment' in this chapter. The defining characteristics of this experimental design are (i) the notion of two groups to be compared (i.e., treatment and control, or A and B) and (ii) the randomized assignment of individuals (e.g., people, cells, computers) to treatment and control groups. Ideally, these individuals will differ on average only in what treatment they receive. Importantly, also typical for these experiments is the assumption that the treatment received by any one individual does not affect the outcomes of any other individuals. That is, it is traditional to assume that there is no *interference* present in the experiment [56, p 19].

Increasingly, however, there is a marked interest in the assessment of treatment effects within networked systems. While experiments on networks are not new, recent advances in technology have in recent years facilitated a sea-change in both their scale and scope. Nowhere is this change more evident than in the social sciences, where experimental social science has undergone a phenomenal transformation, both within traditional areas like economics and sociology and within related areas like marketing. Leveraging the pervasiveness of social media platforms, so-called networked experiments in these areas have explored on previously unthinkable scales topics like the diffusion of knowledge and information, the ability of advertising to influence product adoption, and the spread of political opinions, to name just a few. Insights gained through such experiments can lead to increased understanding of causal mechanisms in social systems, which in turn can be used to tailor interventions like web-based advertising, healthcare monitoring using text messaging, or dissemination of policy statements through Twitter – all aimed at effecting large-scale change at the level of individuals. For a recent summary of much of the literature in this dynamic area, see [15, Table 1].

Naturally, interference cannot realistically be assumed away when doing experiments on networks. As a result, much of what is considered standard in the traditional design of randomized experiments and the corresponding analysis for causal inference does not apply directly in this context. Awareness of interference goes back at least 100 years (e.g., [170]), and its impact on standard theory and methods has been studied previously in certain specific contexts, such as when randomization is done at the level of groups of individuals (e.g., [113]). Nevertheless, the general manner in which interference can manifest itself in networked experiments has led to a flurry of work in network analysis in just the past few years on better characterizing the problems posed by network-based interference (or, simply, 'network interference') and offering appropriate modifications to standard designs and methods of inference. This work appears to be concentrated at the intersection of statistics, economics, and computer science, suggesting the possibility of rich interactions. At the same time, interestingly, the visibility and importance that interference has received through the rise of network-based applications seems in turn to be serving as additional impetus for work on interference in the general literature on causal inference (e.g., [156]).

Our goal in this chapter is to provide a brief but self-contained summary of what would seem to be the key aspects of this emerging and highly active area of research. While a great deal of the related literature is on context-specific problems, experiments, and conclusions, our focus here, in keeping with the spirit of the rest of this monograph, is on the underlying statistics of networked experiments. In Section 4.2 we present necessary notation and background on the notion of counterfactuals and the potential outcomes framework for causal inference. Then, in Section 4.3, we address the topic of causal inference under network interference – first discussing network exposure models and causal estimands under interference, then commenting on design implications, and finally detailing procedures for estimation and quantification of uncertainty. A small illustration is provided in Section 4.4, in the context of organizational behavior.

4.2 Background

Suppose that we have a finite population on which we would like to assess the effectiveness of a treatment, in comparison to the control condition of no treatment. In addition, suppose that there is some notion of a relation between pairs of individuals in our population that is expected to be relevant to the experiment. We represent this scenario through the use of a network graph $G = (V, E)$, where the vertices in V correspond to the N_v individuals in the population, and the edges in E to their relationships. Note that the specification of such a G is effectively a modeling decision and depends on context. For example, in the context of the spread of a virus, an edge $\{i, j\} \in E$ might indicate some notion of 'contact' between individuals i and j during a certain period of time. Alternatively, the same edge might instead mean that individuals i and j are friends on Facebook in the context of an experiment assessing effectiveness of a certain advertising scheme in social media.

Let $z_i = 1$ indicate that individual $i \in V$ received the treatment. We will refer to $\mathbf{z} = (z_1, \ldots, z_{N_v})^T \in \{0, 1\}^{N_v}$ as the treatment assignment vector. By an experimental design we will mean a manner of choosing such a treatment vector. Let $p_{\mathbf{z}} = \mathbb{P}(\mathbf{Z} = \mathbf{z})$ be the probability that treatment assignment $\mathbf{z}$ is generated by the design. Finally, let $\mathcal{O}_i(\mathbf{z})$ denote the outcome for individual i under treatment $\mathbf{z}$. Our treatments here are binary and our outcomes will be assumed continuous, but extensions of what follows can be defined accordingly for other choices.

The goal of evaluating treatment effectiveness based on the observed outcomes $\mathcal{O}_1(\mathbf{z}), \ldots, \mathcal{O}_{N_v}(\mathbf{z})$ is a problem of causal inference. As noted

Table 4.1 Illustration of potential outcomes.

Individual	**'Ideal' World**		**Real World**	
	$Z_i = 0$	$Z_i = 1$	$Z_i = 0$	$Z_i = 1$
1	$\mathcal{O}_1(0)$	$\mathcal{O}_1(1)$	$\mathcal{O}_1(0)$	?
2	$\mathcal{O}_2(0)$	$\mathcal{O}_2(1)$	?	$\mathcal{O}_2(1)$

by Holland [108], the 'fundamental problem of causal inference' is the fact that, although we wish to assess the difference in outcomes of individuals under different treatment options, we are unable to measure multiple outcomes simultaneously on any given individual. The potential outcomes framework – also called the Rubin causal model, or sometimes the Neyman–Rubin causal model [171, 172] – is a framework for causal inference that approaches this problem by first making explicit the notion of all outcomes possible under the experimental design. That is, inclusive of both observed and counterfactual (potential) outcomes. This idea is illustrated in Table 4.1, for $N_v = 2$ individuals. In the ideal world, all four potential outcomes are available to us. However, in the real world, only two of those four are available.

Traditionally, in the standard version of this framework, it is assumed that there is no interference between individuals – that is, $\mathcal{O}_i(\mathbf{z}) = \mathcal{O}_i(z_i)$. This assumption is known in the potential outcomes literature as the stable unit treatment value assumption (SUTVA) and, in fact, was already implicit in our choice of notation in Table 4.1. Under this assumption it becomes meaningful to define the causal effect for individual i (traditionally called the unit-level causal effect) as a function solely of $\mathcal{O}_i(1)$ and $\mathcal{O}_i(0)$. The difference $\mathcal{O}_i(1) - \mathcal{O}_i(0)$ is a popular choice, but in general other functions of these two variables (e.g., the ratio) can be used as well. In turn, various functions of these unit-level causal effects may be of interest in summarizing the effect of treatment at the level of the population in $V = \{1, \ldots, N_v\}$ or subpopulations thereof. A canonical causal estimand of interest is the average treatment effect

$$\tau_{ATE} = \frac{1}{N_v} \sum_{i=1}^{N_v} [\mathcal{O}_i(1) - \mathcal{O}_i(0)] = \overline{\mathcal{O}}(1) - \overline{\mathcal{O}}(0). \tag{4.1}$$

Note that the estimand in (4.1) is a finite-sample population average. Here the perspective is one of design-based inference, in the sense that the randomness in the data is assumed due entirely to the randomization in the experimental design.[1] This randomization is captured through the distribution $p_{\mathbf{z}}$ in our notation above. The standard agenda in this setting is typically as follows: define an unbiased estimator $\hat{\tau}_{ATE}$ of τ_{ATE}, write down the variance $\mathrm{Var}(\hat{\tau}_{ATE})$ of that estimator, formulate an estimate $\hat{\mathrm{V}}\mathrm{ar}(\hat{\tau}_{ATE})$ of that variance, and quantify the uncertainty in $\hat{\tau}_{ATE}$ through, for example, confidence intervals. The details of this agenda depend on the form of $p_{\mathbf{z}}$. Generally, however, while defining $\hat{\tau}_{ATE}$ often is not difficult, estimation of its variance frequently is nontrivial.

For example, consider the classic case of the completely randomized experiment, wherein N_t individuals are assigned to treatment and $N_c = N_v - N_t$ to control, completely at random. Then $p_{\mathbf{z}} = \binom{N_v}{N_t}^{-1}$ and an unbiased estimator of τ_{ATE} is just

$$\hat{\tau}_{ATE}(\mathbf{z}) = \frac{1}{N_v}\sum_{i=1}^{N_v}\left[\frac{z_i\mathcal{O}_i(1)}{N_t/N_v} - \frac{(1-z_i)\mathcal{O}_i(0)}{N_c/N_v}\right]. \tag{4.2}$$

The variance of this estimator is given by

$$\mathrm{Var}\left(\hat{\tau}_{ATE}(\mathbf{Z})\right) = \frac{S_c^2}{N_c} + \frac{S_t^2}{N_t} - \frac{S_{tc}^2}{N_v}, \tag{4.3}$$

where

$$S_c^2 = \frac{1}{N_v - 1}\sum_{i=1}^{N_v}\left[\mathcal{O}_i(0) - \overline{\mathcal{O}}(0)\right]^2 \tag{4.4}$$

and

$$S_t^2 = \frac{1}{N_v - 1}\sum_{i=1}^{N_v}\left[\mathcal{O}_i(1) - \overline{\mathcal{O}}(1)\right]^2 \tag{4.5}$$

[1] In contrast, model-based inference adopts the perspective of the outcomes $\mathcal{O}_i(\cdot)$as being drawn from a super-population, either due to measurement error on specific individuals i or random selection of individuals from a larger population – or both.

are the population-level variances (using the frequent convention of $N - 1$, rather than N, for finite populations) of the individual potential outcomes under control and treatment, respectively, and similarly,

$$S_{tc}^2 = \frac{1}{N_v - 1} \sum_{i=1}^{N_v} [\mathcal{O}_i(1) - \mathcal{O}_i(0) - \tau_{ATE}]^2 \tag{4.6}$$

is the variance of the treatment effects.

Note that the first two terms in (4.3) can be estimated in an unbiased fashion using the sample variances of the outcomes observed for control and treated groups, respectively. Furthermore, if the treatment effects $\mathcal{O}_i(1) - \mathcal{O}_i(0)$ are constant over all individuals i, then the third term in (4.3) is zero, and hence we have an unbiased estimate of $\mathrm{Var}(\hat{\tau}_{ATE})$. In general, however, we cannot estimate the variance in (4.6) based on the observed outcomes alone, since we do not observe both $\mathcal{O}_i(1)$ and $\mathcal{O}_i(0)$ for any of the individuals. As a result, common practice is to use the unbiased estimator derived under the assumption of constant treatment effect (expected to yield a conservative estimate of variance), or to derive estimators based on other or additional assumptions. Confidence intervals typically are based on asymptotic arguments. See [114, Ch 6], for example.

In the context of networks, however, we can in general expect that $\mathcal{O}_i(\mathbf{Z}) \neq \mathcal{O}_i(Z_i)$ – that is, there is interference – and so the stable unit-value treatment assumption is not tenable. Accordingly, much of the recent statistical work on networked experiments has focused on extending the classical paradigm summarized above.

4.3 Causal Inference Under Network Interference

In seeking to extend the classical potential outcomes framework to networked experiments, there is a natural tension between the complexity of the interference in the experiment and the corresponding impact that complexity has on both (i) defining meaningful estimands and (ii) producing estimators thereof and quantifying the uncertainty of those estimators. A natural approach to managing the challenges inherent in this complexity is to place assumptions upon the nature of the interference, to specify an experimental design that is compatible with those assumptions, and to exploit that compatibility to produce tractable and accurate estimators for causal estimands of interest.

4.3.1 Network Exposure Models

In a networked experiment, while the experimenter (ideally) has control over assignment of treatments to individuals, the manner in which any given individual experiences a given treatment assignment as a whole is now assumed to be – at least in principle – a function of the networked system as a whole. As a result, the outcome for individual i arguably is better thought of as a result of the *exposure* of that individual to the full treatment assignment vector $\mathbf{z}$, rather than of only the specific treatment z_i to which that individual was assigned (as under SUTVA). But just how that network exposure manifests at the level of individuals is important. In the worst case, there will be 2^{N_v} possible exposures for each of the N_v individuals, making causal inference impossible.

Intuitively, to avoid this situation, modeling constraints are called for on the extent to which interference from other individuals in the network affect the exposure of a given individual i. Several approaches have been offered in the literature that formalize this idea. Aronow and Samii [17] introduce the notion of so-called exposure mappings, which we adopt here. This approach extends earlier work of Hudgens and Halloran [113] and is analogous to the notion of so-called effective treatments introduced by Manski [146]. The basic idea is to assume that effectively there are only some finite number K of conditions $\{c_1, \ldots, c_K\}$ to which any individual i is exposed. In this way, the complexity of possible exposures is thus reduced. Formally, we say that i is exposed to condition k if $f(\mathbf{z}, \mathbf{x}_i) = c_k$, where $\mathbf{z}$ is the treatment assignment vector, $\mathbf{x}_i$ is a vector of additional information specific to individual i, and f is the exposure mapping.

By way of illustration, note that in the classical setting, wherein it is assumed that there is no interference (i.e., SUTVA), the exposure mapping for each individual i is just $f(\mathbf{z}, \mathbf{x}_i) = z_i$, corresponding to just two conditions c_1 and c_2 (i.e., treated or control). That is, exposure for individual i is determined entirely by the treatment assigned to that individual.

Alternatively, under interference, Aronow and Samii offer a simple, four-level categorization of exposure. Let $\mathbf{A}$ denote the adjacency matrix of the network graph G, and take the vector $\mathbf{x}_i$ to be the ith column of this matrix (i.e., $\mathbf{x}_i = \mathbf{A}_{\cdot i}$). Then define

$$f(\mathbf{z}, \mathbf{A}_{\cdot i}) = \begin{cases} c_{11} \text{ (Direct + Indirect Exposure)}, & z_i I_{\{\mathbf{z}^T \mathbf{A}_{\cdot i} > 0\}} = 1, \\ c_{10} \text{ (Isolated Direct Exposure)}, & z_i I_{\{\mathbf{z}^T \mathbf{A}_{\cdot i} = 0\}} = 1, \\ c_{01} \text{ (Indirect Exposure)}, & (1 - z_i) I_{\{\mathbf{z}^T \mathbf{A}_{\cdot i} > 0\}} = 1, \\ c_{00} \text{ (No Exposure)}, & (1 - z_i) I_{\{\mathbf{z}^T \mathbf{A}_{\cdot i} = 0\}} = 1, \end{cases} \tag{4.7}$$

where the inner product $\mathbf{z}^T \mathbf{A}_{\cdot i}$ is equal to the number of neighbors (or 'peers') of the vertex (or individual) i. Here a double-subscript notation is used to reflect the values of the two indicators z_i and $I_{\{\ldots\}}$. Note that under this definition, each individual i will fall into one and only one of these exposure conditions. As we will see below, this choice of network exposure mapping allows for the definition of various causal estimands capturing to different extents the combination of direct and indirect effects of treatments.

In contrast, Ugander and colleagues [198] suggest several other graph-based notions of network exposure mappings, but for the purposes of their particular choice of estimand it is sufficient to explicitly specify only a subset of the conditions c_k under their mapping. For example, they offer the notion of full-neighborhood exposure (i.e., when an individual i and all neighbors of that individual receive the treatment), as well as variations based on notions of fractional treatment of neighbors and fractional treatment of k-cores.

Finally, we note that while the above approaches involve explicit definitions of network exposure classes, there are also approaches wherein classes are defined implicitly through assumptions on the properties of the network exposure mapping. See, for example, Manski [146] and Leung [136].

4.3.2 Causal Effects Under Network Interference

With the addition of interference to the potential outcomes framework comes the necessity for a corresponding refinement of the notion of causal effects. Recall, for example, the classical definition of an average treatment effect τ_{ATE} in (4.1). A natural generalization of this quantity is

$$\frac{1}{N_v} \sum_{i=1}^{N_v} [\mathcal{O}_i(\mathbf{1}) - \mathcal{O}_i(\mathbf{0})], \tag{4.8}$$

where **1** and **0** are now N_v-length vectors of ones and zeros, respectively. This notion of average treatment effect under network interference is adopted in [198], for example, and corresponds to a comparison of potential outcomes under full treatment (i.e., $\mathbf{z} = \mathbf{1}$) versus full control (i.e., $\mathbf{z} = \mathbf{0}$). In fact, the expression in (4.8) is how average treatment effect is sometimes defined in the classical literature as well, but under the standard stable unit treatment value assumption this reduces to the commonly used expression in (4.1).

The causal estimand in (4.8) represents an aggregation of both direct and indirect effects of the treatment assignment. In many settings it is of interest to disentangle these effects. There have been a number of proposals to date for causal estimands designed to do so, with differences driven largely by context and choice of the exposure model adopted. Note that interference can be viewed alternately as a nuisance, with interest focused on extracting some notion of 'pure' treatment effects (such as in the context of a new putative drug therapy), or instead as the effect of central interest, often representing a mechanism to be exploited (such as from the perspective of social engineering). In the latter context, it is common to see terms like 'peer influence effects' and'spillover effects' used in reference to portions of the treatment effect that are not isolated to the direct contribution of $z_i = 1$ to the outcome $\mathcal{O}_i(\mathbf{z})$.

In the general exposure mapping framework of Aronow and Samii [17], the approach to defining causal estimands is analogous to the classical case, but with the important difference that potential outcomes are defined now at the level of exposure conditions. Specifically, under this framework each individual i has K potential outcomes $\mathcal{O}_i(c_1), \ldots, \mathcal{O}_i(c_K)$. And it is assumed that each individual is exposed to one and only one condition. We use the difference $\mathcal{O}_i(c_k) - \mathcal{O}_i(c_l)$ to represent the causal effect for individual i of exposure condition k versus l. Then, in analogy to (4.1), define

$$\tau(c_k, c_l) = \frac{1}{N_v} \sum_{i=1}^{N_v} [\mathcal{O}_i(c_k) - \mathcal{O}_i(c_l)] = \overline{\mathcal{O}}(c_k) - \overline{\mathcal{O}}(c_l) \tag{4.9}$$

to be the average causal effect of exposure condition k versus l.

To use this framework for disentangling direct and indirect causal effects of treatment requires appropriate definition of $c_1, \ldots, c_K$. Consider again, for example, the exposure mapping function defined in (4.7). A natural set of contrasts is $\tau(c_{01}, c_{00})$, $\tau(c_{10}, c_{00})$, and $\tau(c_{11}, c_{00})$. The first contrast can be interpreted as capturing the average causal effect of treatment over control contributed to an untreated individual through treated neighbors. Note that this effect presumably includes not only that due to treatment of any of the neighbors themselves, but also that due to the treatment of any of the neighbors' neighbors, and so on. Hence, it offers some notion of an overall indirect treatment effect. On the other hand, the second contrast is aimed at capturing some sense of the direct average causal effect of the treatment as applied to an individual i. That is, it is perhaps the closest in spirit to the classical average treatment effect τ_{ATE}. Finally, the third contrast effectively measures the average total treatment effect. Other choices of contrast, of course, are possible as well. Similarly, see Toulis and Kao [197] for related definitions in the context of a different choice of exposure mapping.

4.3.3 Design of Networked Experiments

Throughout our exposition so far in this chapter, the assignment of treatments has been left generic and unspecified. Ultimately, however, the choice of treatment assignment is a key element of network experimental design. At the most fundamental level, treatment assignment is captured in our framework simply through the distribution $p_{\mathbf{z}}$ (i.e., the probability of treatment assignment $\mathbf{z}$). Yet under interference, as we have seen, it is more appropriate to think in terms of overall exposure of individuals to a treatment. Hence, for characterizing a network experimental design, it is useful to have a way of quantifying the manner in which treatment assignment induces exposure under a given network exposure model.

In the context of network exposure mappings, arguably the probability of most immediate interest is now the probability that an individual i is subject to exposure condition k, defined as

$$p_i^e(c_k) = \sum_{\mathbf{Z}} p_{\mathbf{Z}} I_{\{f(\mathbf{Z}, \mathbf{X}_i) = c_k\}}. \tag{4.10}$$

Related probabilities of additional interest are the probability that individuals i and j are both subject to exposure k (i.e., $p_{ij}^e(c_k)$), and the probability that i and j are exposed to k and l, respectively (i.e., $p_{ij}^e(c_k, c_l)$). Definitions of these probabilities are analogous to that in

(4.10). All three of these probabilities will be seen in the next section to play a key role in estimation and uncertainty quantification.

As an illustration, suppose that treatment is assigned to the N_v individuals in a network through Bernoulli random sampling, with probability p, and assume the network exposure model in (4.7). Then, for each individual i, there are four exposure probabilities:

$$\begin{aligned} p_i^e(c_{11}) &= p[1-(1-p)^{d_i}], \\ p_i^e(c_{10}) &= p(1-p)^{d_i}, \\ p_i^e(c_{01}) &= (1-p)[1-(1-p)^{d_i}], \\ p_i^e(c_{00}) &= (1-p)^{d_i+1}, \end{aligned} \tag{4.11}$$

where d_i is the degree of vertex i in the network graph G. Closed-form expressions for the joint exposure probabilities can be obtained similarly but are more involved.

In general, for all individuals i and pairs of individuals i and j, the values of the various exposure probabilities can in principle be recovered from two classes of matrices. Suppose that there are M possible treatment assignments $\mathbf{z}$. Let

$$\mathbf{I}_k = \begin{bmatrix} I_{\{f(\mathbf{Z}_1,\mathbf{X}_1)=c_k\}} & I_{\{f(\mathbf{Z}_2,\mathbf{X}_1)=c_k\}} & \cdots & I_{\{f(\mathbf{Z}_M,\mathbf{X}_1)=c_k\}} \\ I_{\{f(\mathbf{Z}_1,\mathbf{X}_2)=c_k\}} & I_{\{f(\mathbf{Z}_2,\mathbf{X}_2)=c_k\}} & \cdots & I_{\{f(\mathbf{Z}_M,\mathbf{X}_2)=c_k\}} \\ \vdots & \vdots & \ddots & \\ I_{\{f(\mathbf{Z}_1,\mathbf{X}_{N_v})=c_k\}} & I_{\{f(\mathbf{Z}_2,\mathbf{X}_{N_v})=c_k\}} & \cdots & I_{\{f(\mathbf{Z}_M,\mathbf{X}_{N_v})=c_k\}} \end{bmatrix}. \tag{4.12}$$

In addition, let $\mathbf{P} = \operatorname{diag}(p_{\mathbf{Z}_1}, \ldots, p_{\mathbf{Z}_M})$. Then

$$\mathbf{I}_k\mathbf{P}\mathbf{I}_k^T = \begin{bmatrix} p_1^e(c_k) & p_{12}^e(c_k) & \cdots p_{1N_v}^e(c_k) \\ p_{21}^e(c_k) & p_2^e(c_k) & \cdots p_{2N_v}^e(c_k) \\ \vdots & \vdots & \ddots \\ p_{N_v1}^e(c_k) & p_{N_v2}^e(c_k) & \cdots p_{N_v}^e(c_k) \end{bmatrix} \tag{4.13}$$

and

$$\mathbf{I}_k\mathbf{P}\mathbf{I}_l^T = \begin{bmatrix} 0 & p_{12}^e(c_k,c_l) & \cdots p_{1N_v}^e(c_k,c_l) \\ p_{21}^e(c_k,c_l) & 0 & \cdots p_{2N_v}^e(c_k,c_l) \\ \vdots & \vdots & \ddots \\ p_{N_v1}^e(c_k,c_l) & p_{N_v2}^e(c_k,c_l) & \cdots 0 \end{bmatrix}. \tag{4.14}$$

The first matrix, in (4.13), is an $N_v \times N_v$ symmetric matrix, from which we can recover, for a fixed exposure condition c_k, both the individual exposure probabilities $p_i^e(c_k)$ and the joint exposure probabilities $p_{ij}^e(c_k)$, for all individuals $i, j = 1, \ldots, N_v$. The second matrix, in (4.14), is also $N_v \times N_v$ but nonsymmetric, and allows us to recover the joint exposure probabilities $p_{ij}^e(c_k, c_l)$ for all pairs of individuals i and j, for fixed exposure conditions c_k and c_l. Note that diagonal entries in (4.14) are all zero because of the assumption that individuals can only fall into one exposure category.

In practice, it is unlikely that (4.13) and (4.14) will yield closed-form expressions for $p_i^e(c_k)$, $p_{ij}^e(c_k)$, and $p_{ij}^e(c_k, c_l)$. However, Monte Carlo simulation can be used to approximate their values to arbitrary precision. Simulating n draws from $p_{\mathbf{Z}}$, for each of the K exposure conditions, we can then form K matrices $\hat{\mathbf{I}}_k$ of dimension $N_v \times n$, in analogy to the definition of $\mathbf{I}_k$ in (4.12). The estimators $\hat{\mathbf{I}}_k\hat{\mathbf{I}}_k^T/n$ and $\hat{\mathbf{I}}_k\hat{\mathbf{I}}_l^T/n$ are then unbiased for the matrices in (4.13) and (4.14), and converge almost surely by the strong law of large numbers.

A great many practical examples of network experimental design can already be found in the literature, despite the relatively recent emergence of the topic area. Not surprisingly, the details of these examples are often context-specific. See the survey article by Aral [15], for example, and particularly Table 2 therein, for a representative summary of randomization procedures used for networked experiments in the social and economic sciences. In contrast, the formal development and study of network experimental designs from the perspective of statistical theory and methods appears to be in its infancy. Two examples of seminal work in this area are that of Ugander and colleagues [198] and of Toulis and Kao [197].

In [198], the focus is on estimating the notion of average treatment effect defined in (4.8). A so-called graph cluster randomization algorithm is proposed, which is a variation on the type of two-stage hierarchical randomization procedures common in the classical statistical sampling and design literatures. Under this algorithm, the vertex set V of the network graph G is partitioned into, say, n_c clusters (e.g., using any one of the scores of algorithms available for graph partitioning – a.k.a., 'community detection' – in the literature). Randomization is then done at the level of clusters (e.g., Bernoulli assignment or completely randomized assignment), and all vertices within a given cluster receive the same assignment (i.e., either treatment or control). A dynamic programming algorithm is supplied for computing or bounding the relevant individual exposure probabilities. See Hudgens and Halloran [113] as well for an analogous design for causal inference under interference, but proposed without explicit incorporation of the network aspect.

In contrast, in [197] the focus is on the estimation of causal peer influence effects. Defining a notion of network exposure based on whether an individual has exactly, say, m treated neighbors and, similarly, defining a corresponding m-level causal peer influence effect, a randomization procedure is proposed based on sequential randomization. Essentially, in order to ensure estimability of this causal effect, the procedure forces a certain pattern of treatment and control among pairs of individuals i and j that are a distance of two (i.e., two 'hops') away from each other in the network. Specifically, one individual is treated and the other is a control, with each having m treated neighbors, but assigned in such a way that none of the neighbors common to the two individuals are treated. See [197] for details.

There is also work beginning to emerge on the topic of optimal design in networked experiments. See, for example, Basse and Airoldi [23].

4.3.4 *Inference for Causal Effects*

Now consider the problem of inference for causal effects under network interference. Again, we focus on the exposure mapping framework, where estimands of fundamental interest are of two types: (i) the average potential outcomes $\overline{\mathcal{O}}(c_k)$, under exposure conditions c_k, for $k = 1, \ldots, K$, and (ii) the average causal effects $\tau(c_k, c_l) = \overline{\mathcal{O}}(c_k) - \overline{\mathcal{O}}(c_l)$ of exposure conditions k versus l, as defined in (4.9). Note that the observed outcomes corresponding to each estimand of the first type generally will be an unequal-probability sample without replacement of potential outcomes under a given condition (i.e., a sample from the set $\{\mathcal{O}_1(c_k), \ldots, \mathcal{O}_{N_v}(c_k)\}$ under condition c_k). Additionally, the estimands of the second type are all simple linear functions of those in the first. Therefore, assuming that we know or can approximate accurately the relevant exposure probabilities, it is natural to pursue inference following the method of Horvitz and Thompson [111].

The Horvitz–Thompson framework accounts for unequal-probability sampling through the use of inverse probability weighting. Assume all individuals have nonzero exposure probabilities $p_i^e(c_k)$, for all exposure conditions c_k. Then the estimator

$$\hat{\overline{\mathcal{O}}}(c_k) = \frac{1}{N_v} \sum_{i=1}^{N_v} I_{\{f(\mathbf{Z},\mathbf{X}_i)\}=c_k\}} \frac{\mathcal{O}_i(c_k)}{p_i^e(c_k)} \tag{4.15}$$

is well-defined and unbiased for $\overline{\mathcal{O}}(c_k)$. In turn, $\hat{\tau}(c_k, c_l) = \hat{\overline{\mathcal{O}}}(c_k) - \hat{\overline{\mathcal{O}}}(c_l)$ is an unbiased estimator of $\tau(c_k, c_l)$. Furthermore, the variance of the first estimator is

$$\begin{aligned} \text{Var}\left[\hat{\overline{\mathcal{O}}}(c_k)\right] = \frac{1}{N_v^2} \Bigg\{ & \sum_{i=1}^{N_v} p_i^e(c_k)\left[1 - p_i^e(c_k)\right] \left[\frac{\mathcal{O}_i(c_k)}{p_i^e(c_k)}\right]^2 \\ & + \sum_{i=1}^{N_v} \sum_{j \neq i} \left[p_{ij}^e(c_k) - p_i^e(c_k) p_j^e(c_k)\right] \frac{\mathcal{O}_i(c_k)}{p_i^e(c_k)} \frac{\mathcal{O}_j(c_k)}{p_j^e(c_k)} \Bigg\}. \end{aligned} \tag{4.16}$$

and the variance of the second is

$$\text{Var}\Big(\hat{\tau}(c_k, c_l)\Big) = \text{Var}\left[\hat{\overline{\mathcal{O}}}(c_k)\right] + \text{Var}\left[\hat{\overline{\mathcal{O}}}(c_l)\right] - 2\ \text{Cov}\left[\hat{\overline{\mathcal{O}}}(c_k), \hat{\overline{\mathcal{O}}}(c_l)\right], \tag{4.17}$$

where

$$\begin{aligned} \text{Cov}\left[\hat{\overline{\mathcal{O}}}(c_k),\ \hat{\overline{\mathcal{O}}}(c_l)\right] = \frac{1}{N_v^2} \Bigg\{ & \sum_{i=1}^{N_v} \sum_{j \neq i} \frac{\mathcal{O}_i(c_k)}{p_i^e(c_k)} \frac{\mathcal{O}_j(c_l)}{p_j^e(c_l)} \left[p_{ij}^e(c_k, c_l) \right. \\ & \left. - p_i^e(c_k) p_j^e(c_l)\right] - \sum_{i=1}^{N_v} \mathcal{O}_i(c_k) \mathcal{O}_i(c_l) \Bigg\}. \end{aligned} \tag{4.18}$$

As in the classical potential outcomes framework, where interference is absent, so too here the problem of variance estimation must be handled with care. For example, if the joint exposure probabilities $p_{ij}^e(c_k)$ are all strictly greater than zero, the estimator

$$\begin{aligned} \widehat{\text{Var}}\left[\hat{\overline{\mathcal{O}}}(c_k)\right] = \frac{1}{N_v^2} \Bigg\{ & \sum_{i=1}^{N_v} I_{\{f(\mathbf{z},\mathbf{x}_i)=c_k\}} \left[1 - p_i^e(c_k)\right] \left[\frac{\mathcal{O}_i(c_k)}{p_i^e(c_k)}\right]^2 \\ & + \sum_{i=1}^{N_v} \sum_{j \neq i} I_{\{f(\mathbf{Z},\mathbf{X}_i)=c_k\}} I_{\{f(\mathbf{Z},\mathbf{X}_j)=c_k\}} \end{aligned} \tag{4.19}$$

$$\times \frac{p_{ij}^e(c_k) - p_i^e(c_k) p_j^e(c_k)}{p_{ij}^e(c_k)} \frac{\mathcal{O}_i(c_k)}{p_i^e(c_k)} \frac{\mathcal{O}_j(c_k)}{p_j^e(c_k)} \Bigg\} \tag{4.20}$$

is an unbiased estimate of the variance in (4.16). If some of the joint exposure probabilities are equal to zero, then this variance estimator will be biased. However, its bias can be characterized. See Aronow and Samii [17, Sec 5] for details, where a proposal for bias correction may also be found.

In contrast, note that the variance in (4.17) cannot be estimated in an unbiased or consistent manner, due to the last term for the covariance in expression (4.18), which depends upon (unobserved) potential outcomes. However, it is possible to produce a conservatively biased estimator of this variance, in the sense that its expectation is an upper bound on (4.17). See [17, Sec 5].

Finally, as in the classical potential outcomes framework, work to date for producing confidence intervals has rested on asymptotic arguments. Consistency of estimators and coverage guarantees for Wald-based confidence intervals in the exposure mapping framework have been established under appropriate conditions, as the order N_v of the network graph grows. Essentially, these conditions require that (i) the ratios $\mathcal{O}_i(c_k)/p_i^e(c_k)$ of the potential outcomes to their corresponding exposure probabilities are uniformly bounded, (ii) the dependency between treatment exposures across individuals is local and suitably controlled, and (iii) there is a nonzero limiting variance. See [17, Sec 6].

Beyond the exposure mapping framework that underlies the above discussion, other work on estimation of causal effects under network interference has typically leveraged similar concepts and ideas in facing similar challenges. Ugander and colleagues [198], for example, also utilize the Horvitz–Thompson approach in defining their estimators and offer several bounds on the variance of these estimators.

4.4 An Illustration: Influencing Organizational Behavior

We illustrate some of the various concepts and quantities introduced in the preceding sections using the strike dataset of Michael [152] (also analyzed in detail in [60, Ch 7]).

New management took over at a forest products manufacturing facility, and this management team proposed certain changes to the compensation package of the workers. Two union negotiators were responsible for explaining the details of the proposed changes to the employees. The changes were not accepted by the workers, and a strike ensued, which was then followed by a halt in negotiations. At the request of management, who felt that the information about their proposed changes was not being communicated adequately, an outside consultant analyzed the communication structure among 24 relevant actors. When the structure of the corresponding social network was revealed, two specific additional actors were approached and had the changes explained to them, which they then discussed with their colleagues. Within two days the employees requested that their union representatives reopen negotiations and the strike was resolved soon thereafter.

The strategy of explaining the proposed compensation package to a handful of actors in this network can be viewed as an intervention or treatment, in the language of this chapter. And, in hoping that the changes be discussed with other colleagues in the network, it is a treatment designed to exploit the phenomenon of network interference, with the goal of influencing organizational behavior. Since the ultimate goal of the treatment is to convince the employees to accept the proposed changes to the compensation package, the strategy can be viewed as a version of a network persuasion campaign (e.g., [16, 159]). Figure 4.1 shows visual representations of the network describing communication among the employees at the manufacturing facility. An edge between two actors indicates that they communicated at some minimally sufficient level of frequency about the strike. Three subgroups are present in the network: younger, Spanish-speaking employees; younger, English-speaking employees; and older, English-speaking employees.

The two left-hand network visualizations in Figure 4.1 show the pattern of intervention before (top) and after (bottom) involving the outside consultant. Initially, only Sam and Wendle served as negotiators. Afterwards, Bob and Norm were added to the negotiation team. From the perspective of network structure, the choice of these latter two actors is not entirely surprising. Both serve as cut-vertices, in that the removal of either would disconnect the graph. In addition, both have high vertex betweenness centrality.[2] Of the two union representatives,

[2] Betweenness centrality measures are numerical summaries aimed at quantifying the extent to which a vertex (or edge) is located 'between' other pairs of vertices. These centralities are based upon the perspective that 'importance' relates to where a vertex is located with respect to the paths in the network graph. If we picture those paths as the routes by which, say, communication of some sort or another takes place, vertices that sit on many paths are thought likely to be more critical to the communication process. See [127, Ch 4.2.2], for example.

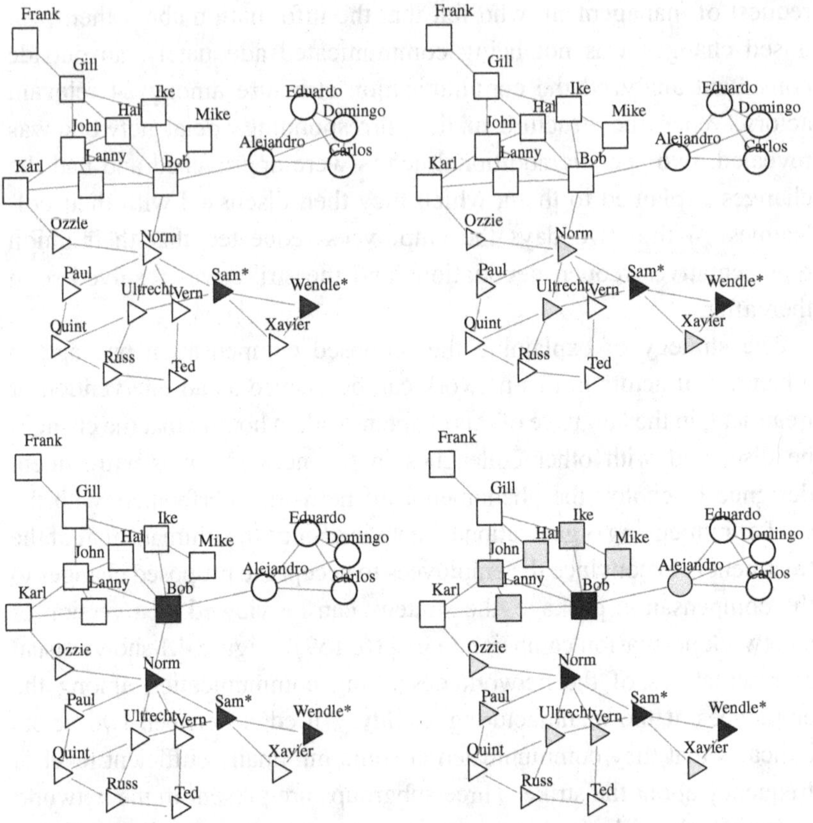

Figure 4.1 The strike network of Michael [152], with three subgroups of employees indicated – younger, Spanish-speaking employees (circular vertices), younger, English-speaking employees (square vertices), and older, English-speaking employees (triangular vertices). *Left:* Those actors serving as negotiators (red) or not (white), under the original communication attempt (top) and after input from an outside consultant (bottom). *Right:* Exposure conditions of each actor under the corresponding intervention (i.e., communication by negotiators), using the definitions in (4.7). Shown are actors that experienced (i) direct plus indirect exposure (i.e., c_{11}, in dark blue), (ii) indirect exposure (i.e., c_{01}, in light blue), and (iii) no exposure (i.e., c_{00}, in white). No actors experienced isolated direct exposure (i.e., c_{10}) under either of these two interventions.

their vertex betweenness values suggest that Sam also plays a nontrivial role in facilitating communication, but that Wendle is not well-situated in this regard. The two right-hand networks in this figure show the exposure conditions of all actors, under the exposure mapping defined in (4.7), for each of the two intervention strategies (i.e., with $N_t = 2$ and 4 people 'treated,' respectively). Clearly the number of actors receiving some combination of direct or indirect exposure to the intervention is substantially increased through use of the four negotiators (i.e., 11 indirect exposures, for 4 treated versus 3 indirect exposures, for 2 treated).

The formation of an appropriate coalition of actors was fundamental to resolving the strike. Suppose, however, that the topology of this network were unknown and, moreover, that the four negotiators were chosen completely at random. Figure 4.2 shows the relevant exposure probabilities for individuals and pairs of individuals under this treatment model, for each of the four exposure conditions. Not surprisingly, with only four actors being 'treated' as negotiators, the exposure probabilities for 'isolated direct exposure' are small (~ 0.10 on average, across actors), while those for 'direct plus isolated exposure' are even smaller still (~ 0.06 on average). In turn, the vast majority of joint exposure probabilities for these two exposure conditions are roughly an order of magnitude smaller again, and many are identically equal to zero. On the other hand, the exposure probabilities for 'indirect exposure' and 'no exposure' are substantially larger, both for individuals and pairs of individuals, and certain implications of network structure are visually apparent as well. For example, we see that in the subgroup of younger, Spanish-speaking employees there is a high probability of no exposure, both individually and pairwise. So too for certain subsets of the other two subgroups. Similarly, we see that Frank has an especially high probability of having no exposure, and therefore joint exposure probabilities for this condition are correspondingly high for most other actors with Frank.

In order to mimic employee reaction to the new compensation plan proposed by management, we adopt a simple model in the spirit of the 'dilated effects' model of Rosenbaum [168]. Suppose that employee reaction is rated on a scale of 1 (nonreceptive) to 10 (completely receptive). Let $\mathcal{O}_i(c_{00}) \equiv 1$ for all employees, indicating a common and standardized level of nonreceptiveness to the proposal prior to any intervention. We then capture a sense of increased receptiveness as

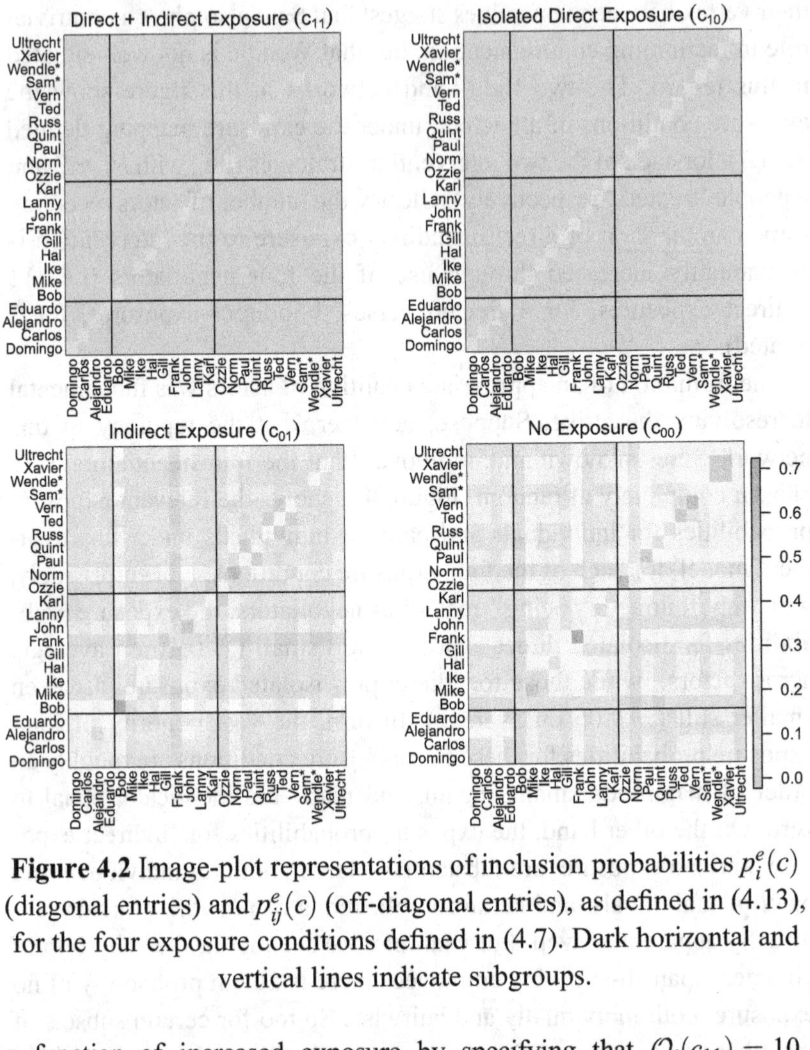

Figure 4.2 Image-plot representations of inclusion probabilities $p_i^e(c)$ (diagonal entries) and $p_{ij}^e(c)$ (off-diagonal entries), as defined in (4.13), for the four exposure conditions defined in (4.7). Dark horizontal and vertical lines indicate subgroups.

a function of increased exposure by specifying that $\mathcal{O}_i(c_{11}) = 10$, $\mathcal{O}_i(c_{10}) = 7$, and $\mathcal{O}_i(c_{01}) = 5$. Thus, the effect of having the compensation package explained directly by management is to increase an employee's receptivity to the proposal from 1 to 7. But among those same employees, if they also are exposed to feedback from others like themselves, their receptivity goes up to a full 10. On the other hand, those employees hearing about the proposal only second-hand see a more moderate rise in receptivity to 5. The corresponding causal effects of interest are $\tau(c_{11}, c_{00}) = 9$, $\tau(c_{10}, c_{00}) = 6$, and $\tau(c_{01}, c_{00}) = 4$.

Monte Carlo approximations of the bias and standard deviation of the Horvitz–Thompson estimators of these estimands, defined through

Table 4.2 Performance of Horvitz–Thompson estimator $\hat{\tau}(\cdot,\cdot)$, defined through (4.15), in estimating causal effects for the strike network. (Based on Monte Carlo simulation of 10,000 trials.)

	Number of Negotiators					
	$N_t = 4$		$N_t = 8$		$N_t = 12$	
Estimand	**Bias**	**S.D.**	**Bias**	**S.D.**	**Bias**	**S.D.**
$\tau(c_{11}, c_{00}) = 9$	-2.6×10^{-2}	8.90	4.1×10^{-2}	2.82	-2.2×10^{-2}	2.75
$\tau(c_{10}, c_{00}) = 6$	2.4×10^{-2}	3.70	5.8×10^{-2}	5.50	-1.2×10^{-1}	10.33
$\tau(c_{01}, c_{00}) = 4$	5.0×10^{-2}	1.61	1.1×10^{-2}	1.55	-2.9×10^{-3}	2.94

(4.15), are shown in Table 4.2, for the case of $N_t = 4, 8$, and 12 'treated' negotiators. As expected, the bias of these estimators is essentially zero for all treatment designs. In contrast, the standard deviation can be substantial in some cases. In particular, it is quite large for estimating total treatment effect (i.e., direct plus indirect, in the first line of the table) when $N_t = 4$, but is noticeably smaller even with $N_t = 8$. At the same time, the standard deviation for estimating the direct effect of treatment (in the second line of the table) starts out moderate in size for $N_t = 4$ but then grows quickly with the number treated. This phenomenon is presumably due to the fact that by $N_t = 8$ or 12, we are treating a large proportion of network nodes, and so the effects of network interference are dominating, making it difficult to obtain estimates of the direct effect of treatment alone. Interestingly, the standard deviation for estimating indirect effects of treatment remain fairly small under all three treatment designs.

Aronow and Samii [17] present the results of an analogous simulation on a much larger network, where the standard deviations are found to be much smaller. Also to be found in that paper are conservative estimates of standard deviation (which are found to be serviceable) and comparisons to other estimators.

4.5 Related Topics

The recently emerged area of networked experiments touches in important ways upon several related literatures. For example, interference in the potential outcomes framework – without any explicit network aspect – has been studied in its own right. This work in fact pre-dates the more recent work on network interference. Interference in this earlier context generally is assumed structured in some useful manner, such as through temporal or spatial adjacency of units or through group membership. Under group-based structural assumptions, interference that maintains only within groups, but not across groups, is known as *partial interference*. The work of Halloran and Struchiner [98] is seminal in framing the nature of the complications that arise under interference and in defining causal effects that attempt to disentangle direct and indirect treatment effects. Key contributions to the corresponding inference problems in this context include Sobel [180], Rosenbaum [169], Hudgens and Halloran [113], and Tchetgen and VanderWeele [194].

Similarly, in parallel with the interest in networked experiments, there has in recent years been a frenzy of interest surrounding the topic of causal inference in network settings where the data are observational, rather than experimental. Much of the interest in this area has centered on the question of whether observed similarity of behaviors and traits of people who share ties (e.g., friendships) is – at least in part – driven by contagion, rather than simply illustrating homophily (i.e., 'birds of a feather flock together'). Seminal work by Christakis and Fowler, presenting empirical evidence suggestive of contagion in traits like obesity [48] and behaviors like smoking [49], has led to a flood of papers exploring causality in the context of networked observational data, from empirical, conceptual, and statistical angles. Unfortunately, one of the main findings resulting from this body of work is that it is in general impossible to identify contagion with such data alone, unless quite strong conditions are made about causal relationships in space and time. Common approaches that use, for example, generalized linear models to establish evidence of contagion (typically by testing whether particular coefficient(s) in the model are zero) have been shown generally to be invalid (e.g., [144]). A careful, self-contained characterization of the challenges inherent in this problem can be found in Shalizi and Thomas [177]. Ultimately, networked experiments – when viable – are in many ways the natural solution to the difficulties encountered in the networked observational context. Although, see Forastiere and colleagues [75] for recent work exploring the observational setting from the perspective of the type of technical developments described herein for the experimental setting.

Finally, the estimation of so-called *peer effects* is by itself an area of substantial interest, particularly in econometrics and related disciplines. Inference in this direction tends to be model-based, with the so-called linear-in-means model serving as a workhorse. A paper of Manski [145] is seminal in this area, wherein notions of endogenous peer effects, exogenous peer effects, and correlated effects are delineated. See also the more recent work of Bramoullé and colleagues [37], who offer an extension of the linear-in-means model in which interactions facilitating peer effects are structured through a social network. Work on semi- and nonparametric extensions of this nature has also recently appeared (e.g., [136]).

Within the topic of networked experiments there are also several additional directions of current research not covered in this chapter which are worth noting. First, while the specific form of 'treatment' adopted here has been that of something applied to individuals (i.e., vertices) in the network, a much broader range of interventions has been explored in practice to date. Other choices include, for example, perturbation of network edges (e.g., turning them 'on' or 'off') or modifying mechanisms by which diffusion takes place along paths. See the survey paper by Aral [15] for extensive discussion and examples in the context of networked experiments in social media. See too the recent work of Eckles and colleagues [68], where the focus is on the reduction of bias in standard estimators through modification of experimental design and integration of design information into analysis.

Second, the focus in this chapter has been squarely on the problem of estimation of treatment effects and, furthermore, restricted to the case where those effects are defined through averages and differences thereof. Yet in the classical literature on potential outcomes, extensions to general causal estimands (e.g., quantiles, Gini coefficients, etc.) have been worked out [114, Ch 20]. Extensions of this nature for networked experiments are presumably in development, as several authors suggest the feasibility of such (e.g., [197]). Furthermore, hypothesis testing for treatment effects under network interference has also seen attention recently, such as in the work of Athey and colleagues [18], who propose a method for calculation of exact p-values for nonsharp null hypotheses.

Lastly, we would be remiss not to mention that one might wish to consider other causal paradigms when studying networked experiments. Here, reflecting the preponderance of the literature to date in this area, we have adopted the potential outcomes framework for causal inference. But also popular, particularly in computer science, is that of graphical models. Ogburn and VanderWeele [156] discuss the connections between these two paradigms specifically when interference is present. See too the paper by Keele [124] for a cogent summary and commentary in this direction.

5
Emerging Topic Areas

5.1 Introduction

Like any active frontier, that between statistics and network analysis is evolving rapidly, and thus is difficult to pin down fully and completely. The topics at the heart of this monograph – network modeling, sampling, and experiments – were chosen in part because they seem to have gathered already a nontrivial critical mass about them. At the same time, however, there are any number of other exciting areas of research emerging at this frontier. Here, in this final chapter, we highlight a few such areas, with the aim of simply sketching in brief some questions of interest and their inherent challenges.

5.2 Dynamic Networks and Multi-Networks

Most complex systems are dynamic in nature. So, realistically, the corresponding network graphs and processes thereon are dynamic as well and, ideally, should be analyzed as such. Friendships (both traditional and online versions) form and dissolve over time. Certain genes may regulate other genes, but only during specific stages of the natural cycle of a cell. And the electrical grid in most countries, while perhaps largely static on the time scale of days or weeks, continues to evolve over the course of years.

At the same time, what we conceptualize as self-contained complex systems, which we describe using individual networks (static or dynamic), often can in turn themselves be viewed as only one of a number of interwoven subsystems within some larger system. As a result, so-called multi-networks – networks consisting of multiple, interconnected layers, with different classes of vertices and/or edges – sometimes can serve as more appropriate representations. For example, friendship is only one designation for a type of relationship between individuals. Similarly, genes are only one functional element participating in a veritable soup of regulatory relationships in a cell, involving numerous other types of elements, such as proteins, metabolites, and lipids. Finally, the electrical grid is just one type of network on which is transported a commodity essential in modern life. Other related networks are the road network and the water network.

In practice, however, the vast majority of network analyses performed to date have been focused on static, single-layer networks. Accordingly, the perspective in this monograph has been focused almost entirely on statistics for such networks. However, both dynamic networks and multi-networks recently have become subjects of intense research interest within the broader network science community. And this interest is beginning to propagate into the field of statistics, although arguably most attention to date within statistics has been given to the former of these two subjects – dynamic networks.

Conceptually, we can think of a dynamic[1] network as a time-indexed graph $G(t) = (V(t), E(t))$, with time t varying in either a discrete or continuous manner over some range of values. Here, $V(t)$ is the set of vertices present in the network at time t, and $E(t)$ the set of values $e_{ij}(t)$, for $i,j \in V(t)$, indicating the presence or absence of edges between those vertices. In practice, however, common constraints on the nature of the measurements we are able to obtain on a complex system actually lead to a number of specific variants of this general formulation.

At one extreme, we may be able to observe with complete accuracy the instantaneous appearance and disappearance of each vertex in time, and similarly the formation and dissolution of edges between the vertices. That is, we may literally observe $G(t)$. And, of course, at the other extreme, we may only be able to obtain a marginal summary of those vertices and edges that appeared at any point during the period of observation, resulting in a static graph G. In between these two extremes lie various other possibilities. For example, we may observe a set of

[1] The adjective 'dynamic' has been applied in the context of networks to describe at least two different aspects of an evolving complex system. Most commonly, it is applied when the edges among a set of vertices – and sometimes the set of vertices itself – are changing as a function of time. Alternatively, it is sometimes used in reference to the attributes of the vertices or edges in a fixed graph G changing in time (e.g., epidemics). The former case may be thought of as referring to dynamics *of* a network, and the latter to dynamics *on* a network. Of course, both types of dynamics may be present together – networks of this type sometimes being referred to as *co-evolving*. Here our focus is on the first of the above cases.

static snapshots of the network, each summarizing the marginal behavior of the network during successive time periods. Such data are sometimes referred to as panel data or longitudinal network data. Alternatively, we may have knowledge of interactions between vertices only up to some finite resolution in time.

Work on dynamic networks in statistics to date appears to have been almost exclusively from the perspective of modeling. Extensions of both latent space models and graphical models have been proposed. Consistent with our emphasis in Chapter 2, we will restrict our comments to the former,[2] and in particular, to work on dynamic versions of stochastic block models. In this context, it is natural to specify a Markov model for the evolution of the underlying latent process associated with class membership. For example, in seminal work, Yang and colleagues [209] specify that the class labels (i.e., specifying to which of the classes $\{1,\ldots,Q\}$ each vertex belongs) evolve according to a discrete-time Markov chain. Alternatively, Xu and Hero [206] allow for temporal evolution of both class labels and the probabilities $\pi_{qr}, q, r \in \{1,\ldots,Q\}$ of connection between vertices of classes q and r. However, as Matias and Miele [148] point out in recent work, the question of identifiability must be handled carefully in this latter case. These last authors – whose primary goal is the inference of class labels over time – provide a characterization of the identifiability problem, propose a variational expectation-maximization (EM) algorithm for estimation of the underlying model parameters, and adopt a version of the integrated classification likelihood in (2.22) for model selection. There also may be found in [148] a relatively complete survey of the related literature.

In comparison, from the perspective of sampling and experiments, there appears to be little statistical work of a foundational nature done so far on dynamic networks – that is, work in the spirit of that described for static networks in Chapters 3 and 4. Certainly, however, there is activity within various domains of application that motivates the need for such work. For example, the analysis of so-called graph streams (i.e., streaming data in the form of graphs) has received substantial interest in the data-mining literature, where sampling has been used in some cases as a component of algorithms aimed at summarization, clustering, and pattern search [4, Sec 4.5]. At the same time, it is not difficult to imagine situations in which network experiments might be conducted with

[2] See [128, Ch 10.5] for a brief discussion of the latter.

multi-stage interventions over time scales that allow for evolution of network connectivity.

Work on statistical modeling, sampling, and design for multi-networks is in its infancy.[3] For example, an extension of stochastic block models has been proposed just recently by Stanley and colleagues [186]. Similarly, there has also been a recent attempt to formalize the notion of sampling from multi-networks, motivated by the problem of monitoring multiple online social networks [178].

5.3 Noisy Networks

In applied network analysis, a common *modus operandi* is to (i) gather basic measurements relevant to the interactions among elements in a system of interest, (ii) construct a network-based representation of that system, with nodes serving as elements and links indicating interactions between pairs of elements, and (iii) use the resulting network graph in further downstream analysis, such as summarizing network structure using a variety of numerical and visual tools. Key here is the point that the process of network analysis usually rests upon some collection of measurements of a more basic nature.

For example, online social networks (e.g., Facebook) are based on the extraction and merging of lists of 'friends' from millions of individual accounts. Similarly, biological networks (e.g., of gene regulatory relationships) are often based on notions of association (e.g., correlation, partial correlation, etc.) among experimental measurements of gene activity levels. Finally, maps of the logical Internet traditionally have been synthesized from the results of surveys in which paths along which information flows are learned through a large set of packet probes (e.g., via traceroute).

Certainly in most applied settings it is widely recognized by practitioners that there is measurement error associated with these and other

[3] In contrast, the one area of statistics that *has* seen nontrivial development to date on multi-networks is the area of network topology inference. The goal in network topology inference is to infer the connectivity of a network from measurements of some combination of vertex and edge attributes. There are a number of canonical versions of this problem. See [127, Ch 7], for example, for background. The version having received the most attention in the statistics literature involves the inference of network edges from vertex measurements alone. Seminal to this topic is the work of Meinshausen and Bühlmann [151], from which a now-large literature has emerged, including work on dynamic networks and, most recently, multi-networks (e.g., [129]).

common types of network constructions. And in many settings the general issue has received substantial attention, such as, for example, in the context of protein–protein interaction networks (e.g., [102]) or social networks (e.g., [12]). But there has been almost no serious attention to date given to the formal probabilistic characterization of the propagation of such error and statistical methods accounting for such propagation. That is, there has been little attention given, at the level of statistical theory and methods, to the problem of noisy networks. Exceptions include the methodology for predicting network topology or attributes with models that explicitly include a component for network noise (e.g., [117, 118]), the 'denoising' of noisy networks (e.g., [45]), and the adaptation of methods for vertex classification using networks observed with errors [161].

An initial contribution to formalizing the notion of, and understanding the nature of, error propagated in analyzing noisy networks has been provided recently in work with Balachandran and Viles [21]. There the focus is on characterizing the distribution of the discrepancy $D = \eta(\hat{G}) - \eta(G)$, in the case where $\eta(\cdot)$ is a subgraph count, where $G = (V, E)$ is the true underlying network and $\hat{G} = (V, \hat{E})$ is a noisy version thereof. The case where the statistic of interest is $\eta(G) = |E|$, the number of edges in G, is studied in some detail. The primary contribution in the paper is to show that in the empirically relevant setting of large graphs with low-rate measurement errors, the distribution of $D_E = |\hat{E}| - |E|$ is well-characterized by a Skellam distribution,[4] when the errors are independent or weakly dependent.

More formally, define the noisy edge set $\hat{E}$ as

$$\hat{E} = \left\{ \{i,j\} \in V^{(2)} : Y_{ij} = 1 \right\},$$

where the marginal distribution of the Y_{ij} is given by

$$Y_{ij} \sim \begin{cases} \text{Bernoulli } (\alpha_{ij}), & \text{if } \{i,\, j\} \in E^c, \\ \text{Bernoulli } (1 - \beta_{ij}), & \text{if } \{i,\, j\} \in E, \end{cases} \tag{5.1}$$

for E^c the set of all non-edges. Here, α_{ij} can be interpreted as the probability of Type I error for a 'test' of $H_0 : \{i,j\} \notin E$ vs. $H_1 : \{i,j\} \in E$, on vertex pair $\{i,j\} \in E^c$. Similarly, β_{ij} can be

[4] The Skellam distribution may be characterized through the difference of two independent Poisson random variables, with means λ_1 and λ_2, respectively.

interpreted as the probability of Type II error for a test of the same hypothesis on vertex pair $\{i,j\} \in E$. Note that the discrepancy in edge counts can be expressed as

$$\begin{aligned} D_E &= |\hat{E}| - |E| \\ &= \sum_{\{i,j\} \in E^c} Y_{ij} - \sum_{\{i,j\} \in E} (1 - Y_{ij}), \end{aligned} \tag{5.2}$$

which is just the number of Type I errors minus the number of Type II errors.

If the network is large enough, the noise level small enough, and the dependency weak enough, then intuitively we might expect each of the sums in (5.2) to behave like Poisson random variables, and their difference like a Skellam random variable. Suppose that the network graph G is large (i.e., N_v tending to infinity) and that the edge noise is homogeneous (i.e., $\alpha_{ij} \equiv \alpha$ and $\beta_{ij} \equiv \beta$), unbiased (in the sense that $\mathbb{E}[|\hat{E}|] = |E|$), and low-rate (in a sense that varies depending on whether G is sparsely or densely connected). Let λ denote the expected number of Type I errors, which, by the assumption of edge-noise unbiasedness, is equal to the number of Type II errors. Then it can be shown [21, Cor 9], under appropriate conditions on the dependency in the noise, that the Kolmogorov–Smirnov distance between the distributions of D_E and a symmetric Skellam random variable, with parameter λ, behaves like

$$d_{KS}(D_E, \text{ Skellam}(\lambda, \lambda)) = O\left(\frac{\text{Var}(D_E) - 2\lambda}{2\lambda}\right). \tag{5.3}$$

While this result is similar in spirit to those found in Chen–Stein theory for Poisson approximations to sums of weakly dependent binary random variables, its proof does not follow directly from such results.

Additional work in this direction is needed. Related results in [21] indicate that an extension of (5.3) to higher-order subgraph counts should be possible, but will be more subtle in nature. Furthermore, simulation results suggest that the analogous Gaussian regime that can be expected to hold at higher noise levels will also likely depend in part on the complexity of the subgraph being counted and the density of the network graph G. More broadly, a parallel line of research is needed for the task of estimating summaries $\eta(G)$ under network noise (i.e., plausibly with something other than plug-in estimates $\eta(\hat{G})$) and for quantifying the uncertainty associated with a given estimator.

5.4 Network Data Objects

Over the past 15–20 years, as the field of network science has exploded with activity, the majority of attention has been focused on the analysis of (usually large) *individual* networks. While it is difficult to envision that the analysis of individual networks will become any less important in the near future, it is likely that in the context of the modern era of 'big data' there will soon be an equal need for the analysis of (possibly large) *collections* of (sub)networks (i.e., collections of network data objects).

We are already seeing evidence of this emerging trend. For example, the analysis of large-scale social networks like Facebook can be facilitated by local analyses, such as through extraction of ego-networks[5] (e.g., [92]). Similarly, the 1000 Functional Connectomes Project,[6] launched a few years ago in imitation of the data-sharing model long-common in computational biology, makes available over 1200 fMRI functional connectivity networks for use and study in the context of computational neuroscience (e.g., [30]). It would seem, therefore, that in the near future, networks of small to moderate size will themselves become standard, high-level data objects.

Faced with databases in which networks are the fundamental unit of data, it will be necessary to have in place the statistical tools to answer such questions as, "What is the 'average' of a collection of networks?" or "Do two collections of networks differ on average? (And, if so, how?)", as well as "What factors (e.g., age, gender, etc.) appear to contribute to differences in networks?" In order to answer these and similar questions, we will require essentially a network-based analogue of the 'Statistics 101' tool box, extending standard tools for scalar and vector data to network data objects.

The extension of such classical tools to network-based datasets, however, is not immediate, since networks are not inherently Euclidean objects. Rather, formally they are combinatorial objects, defined simply through two sets, of vertices and edges, respectively, possibly with an additional corresponding set of weights. Nevertheless, preliminary work in this area demonstrates that networks can be associated with certain natural Euclidean subspaces, and furthermore demonstrates that through a combination of tools from geometry, probability theory, and statistical shape analysis it should be possible to

5 That is, subgraphs formed by each vertex (called 'ego') and its immediate neighbors.
6 http://fcon_1000.projects.nitrc.org/

develop a comprehensive, mathematically rigorous, and computationally feasible framework for producing the desired analogues of classical tools.

For example, Ginestet and colleagues [91] have developed a framework for hypothesis testing when the data objects are weighted networks. Let $G = (V, E, W)$ be a weighted undirected graph, for weights $w_{ij} = w_{ji} \geq 0$, where equality with zero holds if and only if $\{i,j\} \notin E$. Assume G to be simple and connected. Each graph G can then be associated uniquely with its graph Laplacian $L = D(W) - W$, where D is a diagonal matrix of weighted degrees (also called vertex strengths): $D_{jj} = d_j(W) = \Sigma_{i \neq j} w_{ij}$. The notion of a 'space' of networks can then be made precise through the space of such Laplacians, which we denote by $\mathcal{L}$. It can be shown [91, Thm 1] that $\mathcal{L}$ is a submanifold of $\mathbb{R}^{N_v^2}$ of dimension $N_v(N_v - 1)/2$ and, moreover, a convex set in an affine subspace. As a result, a well-defined metric space follows, say $(\mathcal{L}, \rho)$, by equipping $\mathcal{L}$ with the natural choice of Euclidean metric (i.e., where ρ is the Frobenius distance on the space of $N_v \times N_v$ matrices with real-valued entries).

Inference – in particular, nonparametric inference – on such general spaces generally relies on an appropriate definition of means and derivations of asymptotic distributions of sample means. Given a probability measure Q on $(\mathcal{L}, \rho)$, one can define the Fréchet mean (if it exists)

$$\mu = \arg\min_{L' \in \mathcal{L}} \int_{\mathcal{L}} \rho^{\alpha}(L', L) Q(dL), \tag{5.4}$$

where $\alpha \geq 1$ is often taken to be 2. Under independent and identically distributed sampling, the sample Fréchet mean, say $\hat{\mu}_n$, is defined with Q replaced by the empirical measure Q_n (i.e., placing mass $1/n$ at each data point, for n data points). Under appropriate regularity conditions on Q, it may then be shown [91, Thm 3] that

$$n^{1/2}(\hat{\mu}_n - \mu) \rightarrow \text{Normal}(0, \Sigma), \tag{5.5}$$

where μ and $\hat{\mu}_n$ are to be understood as being in half-vectorization (i.e., vech) form, $\Sigma = \text{Cov}[\text{vech}(L)]$, and '$\rightarrow$' denotes convergence in distribution as n tends to infinity.

In analogy to many inference tools taught in a traditional 'Statistics 101' course, results like that above can serve as the foundation for the development of tools in a number of directions. For example, various tests of hypotheses involving μ may be built off this result. See [91] for a handful of such tests, as well as an illustration of their application to the 1000 Functional Connectomes dataset.

Other results of this type are beginning to emerge. These include additional work on asymptotics for network sample means [193] and regression modeling with a network response variable, where for the latter there have been both frequentist [55] and Bayesian [66] proposals. Earlier efforts in this space have focused on the specific case of trees. Contributions of this nature include work on central limit theorems in the space of phylogenetic trees [22, 29] and work by Marron and colleagues [20, 202] in the context of so-called object-oriented data analysis with trees.

Appendix

Background on Graphs

We review here some basic terminology and concepts on graphs, introduce certain important connections between graphs and matrix algebra, and briefly visit the topic of graph data structures and algorithms. Our intent is to supply readers from a primarily statistical background a minimal level of exposure to the mathematical and computational aspects of networks, sufficient to comfortably engage with the topics of this book and the related literature. A more thorough introduction to the topic of graph theory may be found in any of a number of introductory textbooks, such as those by Bollobás [31], Diestel [63], or Gross and Yellen [96]. Details on graph data structures and algorithms are in many computer science algorithms texts. See the text by Cormen, Leiserson, Rivest, and Stein [54], for example.

A.1 Basic Definitions and Concepts

Formally, a *graph* $G = (V, E)$ is a mathematical structure consisting of a set V of *vertices* (also commonly called *nodes*) and a set E of *edges* (also commonly called *links*), where elements of E are unordered pairs $\{u, v\}$ of distinct vertices $u, v \in V$. The number of vertices $N_v = |V|$ and the number of edges $N_e = |E|$ are sometimes called the *order* and *size* of the graph G, respectively. Often, and without loss of generality,[1] the vertices are labeled simply with the integers $1, \ldots, N_v$, and the edges analogously. A graph $H = (V_H, E_H)$ is a *subgraph* of another graph $G = (V_G, E_G)$ if $V_H \subseteq V_G$ and $E_H \subseteq E_G$. An *induced subgraph* of G is a subgraph $G' = (V', E')$, where $V' \subseteq V$ is a prespecified subset of

[1] Technically, a graph G is unique only up to relabelings of its vertices and edges that leave the structure unchanged. Two graphs that are equivalent in this sense are called *isomorphic*.

vertices and $E' \subseteq E$ is the collection of edges to be found in G among that subset of vertices.

As defined, a graph has no edges for which both ends connect to a single vertex (called *loops*) and no pairs of vertices with more than one edge between them (called *multi-edges*). A graph with either of these properties is called a *multi-graph*. A graph is referred to as *simple*, and its edges as *proper*, when there are no loops or multi-edges.

A graph G for which each edge in E has an ordering to its vertices (i.e., so that $\{u, v\}$ is distinct from $\{v, u\}$, for $u, v \in V$) is called a *directed graph* or *digraph*. Such edges are called *directed edges* or *arcs*, with the direction of an arc $\{u, v\}$ read from left to right, from the *tail* u to the *head* v. Note that there is a natural extension of digraphs to *multi-digraphs*, where multiple arcs (i.e., *multi-arcs*) share the same head and tail. Note too, however, that digraphs may have two arcs between a pair of vertices without their being multi-arcs if the vertices play opposite roles of head and tail for the respective arcs. In this case, the two arcs are said to be *mutual*.

It is necessary to have a language for discussing the connectivity of a graph. One of the most basic notions of connectivity is that of adjacency. Two vertices $u, v \in V$ are said to be *adjacent* if joined by an edge in E. Similarly, two edges $e_1, e_2 \in E$ are adjacent if joined by a common endpoint in V. A vertex $v \in V$ is *incident* on an edge $e \in E$ if v is an endpoint of e.

From this follows the notion of the *degree* of a vertex v, say d_v, defined as the number of edges incident on v. The *degree sequence* of a graph G is the sequence formed by arranging the vertex degrees d_v in nondecreasing order. The sum of the elements of the degree sequence is equal to twice the number of edges in the graph (i.e., twice the size of the graph). Note that for digraphs, vertex degree is replaced by *in-degree* (i.e., d_v^{in}) and *out-degree* (i.e., d_v^{out}), which count the number of edges pointing in towards and out from a vertex, respectively. Hence, digraphs have both an in-degree sequence and an out-degree sequence.

It is also useful to be able to discuss the concept of movement about a graph. For example, a *walk* on a graph G, from v_0 to v_l, is an alternating sequence $\{v_0, e_1, v_1, e_2, \ldots, v_{l-1}, e_l, v_l\}$, where the endpoints of e_i are $\{v_{i-1}, v_i\}$. The *length* of this walk is said to be l. Refinements of a walk include *trails*, which are walks without repeated edges, and *paths*, which are trails without repeated vertices. A trail for which the beginning and ending vertices are the same is called a *circuit*. Similarly, a walk of length at least three, for which the beginning and ending vertices are the same, but for which all other vertices are distinct from each other, is called a *cycle*. Graphs containing no cycles are called *acyclic*. In a digraph, these notions generalize naturally. For example, a *directed walk* from v_0 to v_l proceeds from tail to head along arcs between v_0 and v_l.

A vertex v in a graph G is said to be *reachable* from another vertex u if there exists a walk from u to v. The graph G is said to be *connected* if every vertex is reachable from every other. A *component* of a graph is a maximally connected subgraph. That is, it is a connected subgraph of G for which the addition of any other remaining vertex in V would ruin the property of connectivity. For a digraph, there are two variations of the concept of connectedness. A digraph G is *weakly connected* if its underlying graph (i.e., the result of stripping away the labels 'tail' and 'head' from G) is connected. It is called *strongly connected* if every vertex v is reachable from every u by a directed walk.

A common notion of *distance* between vertices on a graph is defined as the length of the shortest path(s) between the vertices (which commonly is set equal to infinity if no such path exists). This distance is often referred to as *geodesic distance*, with 'geodesic' being another name for shortest paths. The value of the longest distance in a graph is called the *diameter* of the graph.

Finally, it is not uncommon to equip (or 'decorate') a graph G with auxiliary numerical values on its vertices, edges, or both. For example, edges $e \in E$ are often accompanied by *edge weights*. In fact, extending the notion of edge weights to all pairs of vertices, the edge set E itself can be represented through a set $\{w_e\}$ of such weights (i.e., $w_e = 1$ if $e \in E$ and 0 if $e \notin E$). When edges are weighted, the corresponding length of a walk (trail, path, etc.) is measured as the sum of the values of the weights along the edges traversed in the walk. The notion of distance generalizes accordingly. These concepts extend naturally to digraphs.

Similarly, graph labelings may be used in representing a multi-graph as a decorated graph. Specifically, given a multi-graph, we can define a graph G having the same vertex set V and having an edge set E such that distinct elements $u, v \in V$ have an edge between them if there is at least one multi-edge between them in the multi-graph. Then, equip each vertex $v \in V$ with a label denoting the number of loops possessed by v in the multi-graph, and similarly, equip each edge with the number of multi-edges it represents.

Of course, from a statistical perspective, a particularly common source of labels for graph vertices and edges will be in the form of measurements of functions or processes on a given graph. From a network-centric perspective, these other data typically are referred to as *attributes* of the graph. Vertex attributes are variables indexed by vertices, while edge attributes similarly are variables indexed by adjacent vertex pairs. In both cases, attributes may be of either discrete or continuous type. Examples of vertex attributes include the gender of actors in a social network or the voltage potential levels in the brain measured at electrodes in an electrocorticogram (ECoG) grid. Similarly, examples of edge attributes include whether two countries have a friendly or antagonistic political relationship or the average time necessary during a given hour of the day for trains to run from one to station in a subway network to the next.

A.2 Graphs and Matrix Algebra

From a statistical perspective, in a field where vector and matrix representations and linear algebra are at the classical core, and still persist strongly throughout our literature, it is particularly useful to be able to characterize a graph G and certain aspects of its structure using matrices and matrix algebra. Our ability to do so in a rigorous manner derives from a blending of graph theory with matrix algebra, represented formally by the area known as algebraic graph theory. While this area has been developed extensively, understanding just a handful of elements from it can, for a statistician, go a long way.

Most basic is the *adjacency matrix*, which captures the fundamental connectivity of a graph G. This matrix is an $N_v \times N_v$ binary, symmetric matrix, say $\mathbf{A}$, with entries

$$A_{ij} = \begin{cases} 1, & \text{if}\{i,j\} \in E, \\ 0, & \text{otherwise.} \end{cases} \tag{A.1}$$

In words, $\mathbf{A}$ is nonzero for entries whose row–column indices correspond to vertices in G joined by an edge, and zero for those that do not.

The adjacency matrix is, not surprisingly, the standard way in which the connectivity of a graph G enters into statistical modeling. It is useful not only for storing connectivity information, but also in that certain operations on $\mathbf{A}$ yield additional information concerning G. For example, the row sum $A_{i+} = \sum_j A_{ij}$ is simply equal to the degree d_i of vertex i. Note that, by symmetry, $A_{i+} = A_{+i}$. Furthermore, if we let $\mathbf{A}^r$ denote the rth power of

$\mathbf{A}$, then the entry A^r_{ij} yields the number of walks of length r between i and j on G. Finally, there are many interesting and useful relations involving the eigenvalues of G. For example, it can be shown that G is a regular graph if and only if the maximum degree d_{max} of G is an eigenvalue of $\mathbf{A}$.

An adjacency matrix may also be defined for digraphs, by adjusting the definition in (A.1) so that $A_{ij} = 1$ if $\{i,j\} \in E$ represents a directed edge from i to j. Of course, $\mathbf{A}$ is now no longer symmetric. However, it still contains similarly useful additional information. For example, $A_{i+} = d_i^{out}$ and $A_{+j} = d_j^{in}$.

The $N_v \times N_v$ matrix $\mathbf{L} = \mathbf{D} - \mathbf{A}$ is also important in its own right and is called the *Laplacian* of the graph G. Motivation for the term Laplacian, in analogy to the Laplacian from multivariable calculus (i.e., the sum of second partial derivatives of a function), may be found in the fact that, for a real-valued vector $\mathbf{x} \in \mathbb{R}^{N_v}$, we have

$$\mathbf{x}^T \mathbf{L} \mathbf{x} = \sum_{\{i,j\} \in E} (x_i - x_j)^2. \tag{A.2}$$

The closer this value is to zero, the more similar are the elements of $\mathbf{x}$ at adjacent vertices in V. Hence, the Laplacian is useful in providing, through (A.2), some sense of the 'smoothness' of functions on a graph G, with respect to the connectivity of G.

The properties of $\mathbf{L}$, particularly the properties of its eigenvalues and eigenvectors, have much to say about the structure of G. Since $\mathbf{L}$ can be shown to be a positive semi-definite matrix, the eigenvalues are all non-negative. And because $\mathbf{L1} = \mathbf{0}$, where $\mathbf{1}$ and $\mathbf{0}$ are $n_v \times 1$ vectors of ones and zeros, respectively, the smallest eigenvalue λ_1 of $\mathbf{L}$ is equal to zero. The second smallest eigenvalue, λ_2, is typically nontrivial, and arguably the most important of all the eigenvalues. For example, roughly speaking, the larger λ_2 is, the more 'connected' G is, and the more difficult it is to separate G into disconnected subgraphs by selectively eliminating some small number of edges. As a result, the Laplacian has played a key role in, for example, the development of methods of community detection (or graph partitioning) in networks. See [51] for an overview of the graph Laplacian and its properties.

A.3 Graph Data Structures and Algorithms

Central to the transition from graphs as purely mathematical objects to graphs as practical tools for use in network analysis are data structures and algorithms for graphs. The study of graph data structures and algorithms is basic to the field of computer science, where much effort has gone into the development of efficient methods for the storage and manipulation of a graph, as well as methods for computing various characteristics of – and answering different questions about – graphs. For the statistician, from a practical point of view, not only is it important to be aware of the various manners in which network data we might receive are structured, it is critical to have at least a passing familiarity with the impact that algorithmic complexity can have on computational efficiency of new methodologies we might propose.

A.3.1 Data Structures

There are two common data structures for representing a graph G. The first is the $N_v \times N_v$ adjacency matrix $\mathbf{A}$ defined previously. This choice is often a natural one, given that matrices are fundamental data objects in most standard programming and software environments. However, if the graph is particularly large, and especially if the graph is sparse (e.g., if $N_e \sim N_v$), then it can be preferable to use a collection of *adjacency lists*. This is because the adjacency matrix in such cases will be both large and filled primarily with zeros, due to the fact that it explicitly represents both present and absent edges, while adjacency lists, on the other hand, store only the information on edges that are present. Specifically, an adjacency-list representation of a graph G is simply an array of size N_v, ordered with respect to the ordering of the vertices in V, each element of which is a list, where the ith list contains the set of all vertices j for which there is an edge from i to j. A variation on this idea is an *edge list*, a simple two-column list of all vertex pairs that are joined by an edge.

The sum of the lengths of the adjacency lists will be N_e for a directed graph and $2N_e$ for an undirected graph. Therefore, the total amount of computer memory space required for storing a graph in an adjacency list representation is only $O(N_v + N_e)$. When G is sparse, this memory requirement will be much less than the value $O(N_v^2)$ associated with storage of G through an adjacency matrix. On the other hand, when G is dense (i.e., $N_e \sim N_v^2$), the memory requirements for the two methods will be comparable. In addition, the simplicity of the adjacency matrix representation may sometimes be felt to outweigh any memory disadvantages, especially for smaller graphs.[2]

Note that for either adjacency-matrix or adjacency-list representations, it is easy to store annotated graphs in a similar fashion. For example, an edge weight w_e, for $e = \{i,j\} \in E$, may be stored in the (i,j) th location of the adjacency matrix, or in addition to the value j (or i) in the list for i (or j), in an adjacency-list representation.

A.3.2 Algorithms

There are numerous questions we might wish to ask in regard to a given graph. Some are quite simple and their answers may be obtained in a straightforward manner directly from the data structures just discussed. For example, are vertices i and j linked by an edge? What is the degree of vertex i? For many other questions, however, more work may be required, but it is usually still feasible to obtain an answer in a reasonably efficient manner. For example, what is the shortest path(s) between vertices i and j? How many connected components does the graph have? And, for a directed graph, does it have cycles or is it acyclic? Finally, for certain questions, there is likely to be no efficient algorithm. An example of such a question is the one that asks for a maximal clique in a given graph.

Thus, algorithmic complexity is frequently an important issue in network analysis. Computational complexity theory distinguishes between 'tractable' and 'intractable' problems by breaking them into two groups – those that are solvable with an algorithm that runs in polynomial time, and those that are not. A polynomial-time algorithm with n inputs will run in time $O(n^p)$ for some $p > 0$. If an algorithm is not polynomial for any choice of p, it is super-polynomial. For example, combinatorially exhaustive algorithms are often exponential, in the sense that they run in $O(a^n)$ time for some $a > 1$. Except for settings with very small n, it is unlikely to be feasible in practice to wait for super-polynomial algorithms to complete; the same is effectively true even for polynomial algorithms having sufficiently large p, with respect to n.

[2] Ultimately, a comparison of the two methods of data representation for graphs can be more subtle than we have described. For example, sparse matrix storage methods allow one to effectively store the adjacency matrix in significantly less than $O(N_v^2)$ space. In addition, there can be important interactions between data representation and algorithm design that impact computational efficiency.

In the case of graph algorithms, the complexity of an algorithm usually is a function of both N_v and N_e. Ideally, one would like the running time of such algorithms to scale linearly in these values (i.e., with $p = 1$), while also keeping memory requirements under control. In practice, for graphs of any reasonably large dimension, a cubic algorithm (i.e., with $p = 3$) is generally pushing the upper limits for what may be computed in an acceptable amount of time. In the case of massive network graphs, however, often nothing less efficient than linear running time is feasible. The design of efficient algorithms is frequently nontrivial, with a more naive, inefficient algorithm being improved only through clever use of storage, indexing, or redundancies in the quantities to be computed.

Consider, for example, the notion of a 'search' on a graph, which we can picture as a process of moving outward from a given vertex towards the other vertices in the graph, seeking to fulfill a given criterion. The two most basic search algorithms, *breadth-first search* (BFS) and *depth-first search* (DFS), each seek to 'discover' the set of all vertices $j \in V$ that are reachable from a given source vertex i_s. Both algorithms run in $O(N_v + N_e)$ time, and differ from each other primarily in the manner in which they choose to discover other vertices. The BFS algorithm works outward from i_s, first discovering vertices adjacent to i_s (i.e., one 'hop' away), then continuing to vertices two hops away, then three hops away, and so on, until all reachable vertices are discovered. The output of the algorithm is a tree, rooted at i_s and organized so that the path from i_s to a reachable vertex j in the tree corresponds to the shortest path from i_s to j in G. Conversely, the DFS algorithm, as its name suggests, instead proceeds from i_s by delving as deeply into G as possible from the first adjacent vertex to i_s, after which it iteratively backtracks to the most recently discovered vertex j for which there are undiscovered edges to be explored.

The usefulness of these two search algorithms, and the choice of which to use when, derives from the context of the actual task to be performed. For example, the BFS algorithm is at the core of standard algorithms requiring shortest-path information, like Prim's algorithm for producing a minimum spanning tree[3] and Dijkstra's algorithm for finding all shortest paths from a single source i_s in a directed graph.

[3] A minimum spanning tree $T \in E$ is a tree that spans the full set of vertices V and for whom the sum of edge weights $W(T) = \sum_{e \in T} w_e$ is minimized.

The latter algorithms can be implemented to run in $O(N_e + N_v \log N_v)$ and $O(N_v^2 \log N_v + N_v N_e)$ time, respectively. The DFS algorithm is often a subroutine in a larger algorithm, such as the 'topological sort algorithm,' which can be used to determine whether a directed graph G is acyclic or not, and algorithms for decomposing G into its strongly connected components, all of which can be implemented in linear time.

Interestingly, there are many problems for which it is unknown whether there even exists a polynomial-time algorithm to solve them. The study of such problems has been a deep and fundamental endeavor of theoretical computer science since the early 1970s. Within this area there is a formal framework for classifying problems by how difficult they *may* be. Decision problems (i.e., problems requiring a 'yes' or 'no' answer) that are shown to be *NP-hard* or *NP-complete* are considered quite likely to be intractable.[4] For optimization problems, wherein we seek to find a solution maximizing or minimizing a given objective function, the problem is referred to by these same terms if the associated decision problem of confirming or rejecting a candidate solution as optimal can be categorized as such. In the context of network graphs, the problem of finding a maximal clique in a graph is an example of an *NP*-complete optimization problem.

[4] The designations *NP*-hard and *NP*-complete refer to two different but related notions of the difficulty of a problem. See, for example, the text by Cormen, Leiserson, Rivest, and Stein [54, Ch 34].

References

[1] Abbe, E., and Sandon, C. 2015. Community detection in general stochastic block models: Fundamental limits and efficient algorithms for recovery. *56th IEEE Annual Symposium on Foundations of Computer Science (FOCS)*, 670–688.

[2] Achlioptas, D., Clauset, A., Kempe, D., and Moore, C. 2005. On the bias of trace-route sampling. *Proceedings of the 37th Annual ACM Symposium on Theory of Computing*, 694–703.

[3] Ahmed, N.K., Neville, J., and Kompella, R. 2010. Reconsidering the foundations of network sampling. *Proceedings of the 2nd Workshop on Information in Networks*.

[4] Ahmed, N.K., Neville, J., and Kompella, R. 2014. Network sampling: From static to streaming graphs. *ACM Transactions on Knowledge Discovery from Data (TKDD)*, **8**(2), 7.

[5] Aicher, C., Jacobs, A.Z., and Clauset, A. 2014. Learning latent block structure in weighted networks. *Journal of Complex Networks*, cnu026.

[6] Airoldi, E.M., Blei, D.M., Fienberg, S.E., and Xing, E.P. 2008. Mixed membership stochastic blockmodels. *Journal of Machine Learning Research*, **9**(Sep), 1981–2014.

[7] Airoldi, E.M., Choi, D.S., and Wolfe, P.J. 2011. Confidence sets for network structure. *Statistical Analysis and Data Mining*, **4**(5), 461–469.

[8] Airoldi, E.M., Costa, T.B., and Chan, S.H. 2013. Stochastic blockmodel approximation of a graphon: Theory and consistent estimation. *Advances in Neural Information Processing Systems*, 692–700.

[9] Aldous, D. 1985. Exchangeability and related topics. *École d'Été de Probabilités de Saint-Flour XIII1983*, 1–198.

[10] Allman, E.S., Matias, C., and Rhodes, J.A. 2009. Identifiability of parameters in latent structure models with many observed variables. *Annals of Statistics*, **37**(6A), 3099–3132.

[11] Allman, E.S., Matias, C., and Rhodes, J.A. 2011. Parameter identifiability in a class of random graph mixture models. *Journal of Statistical Planning and Inference*, **141**(5), 1719–1736.

[12] Almquist, Z.W. 2012. Random errors in egocentric networks. *Social Networks*, **34**(4), 493–505.

[13] Ambroise, C., and Matias, C. 2012. New consistent and asymptotically normal parameter estimates for random-graph mixture models. *Journal of the Royal Statistical Society: Series B (Statistical Methodology)*, **74**(1), 3–35.

[14] Amini, A.A., Chen, A., Bickel, P.J., and Levina, E. 2013. Pseudo-likelihood methods for community detection in large sparse networks. *Annals of Statistics*, **41**(4), 2097–2122.

[15] Aral, S. 2016. Networked experiments: A review of methods and innovations. In: Bramoulle, Y., Galeotti, A., and Rogers, B. (eds), *The Oxford Handbook of the Economics of Networks*. Oxford: Oxford University Press.

[16] Aral, S., and Walker, D. 2011. Creating social contagion through viral product design: A randomized trial of peer influence in networks. *Management Science*, **57**(9), 1623–1639.

[17] Aronow, P.M., and Samii, C. 2013. Estimating average causal effects under interference between units. *arXiv preprint arXiv:1305.6156*.

[18] Athey, S., Eckles, D., and Imbens, G.W. 2017. Exact p-values for network interference. *Journal of the American Statistical Association* (in press).

[19] Austin, T. 2008. On exchangeable random variables and the statistics of large graphs and hypergraphs. *Probability Surveys*, **5**(1), 80–145.

[20] Aydin, B., Pataki, G., Wang, H., Bullitt, E., and Marron, J.S. 2009. A principal component analysis for trees. *Annals of Applied Statistics*, **3** (4), 1597–1615.

[21] Balachandran, P., Kolaczyk, E.D., and Viles, W.D. 2014. On the propagation of low-rate measurement error to subgraph counts in large networks. *arXiv preprint arXiv:1409.5640*.

[22] Barden, D., Le, H., and Owen, M. 2013. Central limit theorems for Fréchet means in the space of phylogenetic trees. *Electronic Journal of Probability*, **18**(25), 1–25.

[23] Basse, G.W., and Airoldi, E.M. 2016. Optimal model-assisted design of experiments for network correlated outcomes suggests new notions of network balance. *arXiv preprint arXiv:1507.00803*.

[24] Bickel, P.J., and Chen, A. 2009. A nonparametric view of network models and Newman–Girvan and other modularities. *Proceedings of the National Academy of Sciences*, **106**(50), 21068–21073.

[25] Bickel, P.J., and Sarkar, P. 2016. Hypothesis testing for automated community detection in networks. *Journal of the Royal Statistical Society: Series B (Statistical Methodology)*, **78**(1), 253–273.

[26] Bickel, P.J., Chen, A., and Levina, E. 2011. The method of moments and degree distributions for network models. *Annals of Statistics*, **39**(5), 2280–2301.

[27] Bickel, P.J., Choi, D.S., Chang, X., and Zhang, H. 2013. Asymptotic normality of maximum likelihood and its variational approximation for stochastic blockmodels. *Annals of Statistics*, **41**(4), 1922–1943.

[28] Biernacki, C., Celeux, G., and Govaert, G. 2000. Assessing a mixture model for clustering with the integrated completed likelihood. *IEEE Transactions on Pattern Analysis and Machine Intelligence*, **22**(7), 719–725.

[29] Billera, L.J., Holmes, S., and Vogtmann, K. 2001. Geometry of the space of phylogenetic trees. *Advances in Applied Mathematics*, **27**(4), 733–767.

[30] Biswal, B.B., Mennes, M., Zuo, X.-N., Gohel, S., Kelly, C., Smith, S.M., *et al.* 2010. Toward discovery science of human brain function. *Proceedings of the National Academy of Sciences*, **107**(10), 4734–4739.

[31] Bollobás, B. 1998. *Modern Graph Theory*. New York: Springer.

[32] Bollobás, B. 2001. *Random Graphs*, 2nd edn. New York: Cambridge University Press.

[33] Bollobás, B., and Riordan, O. 2011. Sparse graphs: Metrics and random models. *Random Structures & Algorithms*, **39**(1), 1–38.

[34] Bollobás, B., Janson, S., and Riordan, O. 2007. The phase transition in inhomogeneous random graphs. *Random Structures & Algorithms*, **31**(1), 3–122.

[35] Borgs, C., Chayes, J., Lovász, L., Sós, V.T., Szegedy, B., and Vesztergombi, K. 2006. Graph limits and parameter testing. *Proceedings of the 38th Annual ACM Symposium on Theory of Computing*, 261–270.

[36] Borgs, C., Chayes, J.T., Lovász, L., Sós, V.T., and Vesztergombi, K. 2008. Convergent sequences of dense graphs I: Subgraph frequencies, metric properties and testing. *Advances in Mathematics*, **219**(6), 1801–1851.

[37] Bramoullé, Y., Djebbari, H., and Fortin, B. 2009. Identification of peer effects through social networks. *Journal of Econometrics*, **150**(1), 41–55.

[38] Brandes, U., and Pich, C. 2007. Centrality estimation in large networks. *International Journal of Bifurcation and Chaos*, **17**(7), 2303–2318.

[39] Brault, V., and Mariadassou, M. 2015. Co-clustering through latent bloc model: A review. *Journal de la Société Française de Statistique*, **156**(3), 120–139.

[40] Bunge, J., and Fitzpatrick, M. 1993. Estimating the number of species: A review. *Journal of the American Statistical Association*, **88**(421), 364–373.

[41] Caron, F., and Fox, E.B. 2014. Sparse graphs using exchangeable random measures. *arXiv preprint arXiv:1401.1137*.

[42] Celisse, A., Daudin, J.-J., and Pierre, L. 2012. Consistency of maximum-likelihood and variational estimators in the stochastic block model. *Electronic Journal of Statistics*, **6**, 1847–1899.

[43] Chan, S.H., and Airoldi, E.M. 2014. A consistent histogram estimator for exchangeable graph models. *Proceedings of the 31st International Conference on Machine Learning*, 208–216.

[44] Chandrasekhar, A.G., and Jackson, M.O. 2014. *Tractable and consistent random graph models*. Technical Report of the National Bureau of Economic Research.

[45] Chatterjee, S. 2015. Matrix estimation by universal singular value thresholding. *Annals of Statistics*, **43**(1), 177–214.

[46] Chatterjee, S., and Diaconis, P. 2013. Estimating and understanding exponential random graph models. *Annals of Statistics*, **41**(5), 2428–2461.

[47] Choi, D.S., Wolfe, P.J., and Airoldi, E.M. 2012. Stochastic blockmodels with a growing number of classes. *Biometrika*, **99**(2), 273–284.

[48] Christakis, N.A., and Fowler, J.H. 2007. The spread of obesity in a large social network over 32 years. *New England Journal of Medicine*, **357**(4), 370–379.

[49] Christakis, N.A., and Fowler, J.H. 2008. The collective dynamics of smoking in a large social network. *New England Journal of Medicine*, **358**(21), 2249–2258.

[50] Chung, F., and Lu, L. 2006. *Complex Graphs and Networks*. American Mathematical Society.

[51] Chung, F.R.K. 1997. *Spectral Graph Theory*. American Mathematical Society.

[52] Clauset, A., and Moore, C. 2005. Accuracy and scaling phenomena in Internet mapping. *Physical Review Letters*, **94**(1), 18701.

[53] Clauset, A., Moore, C., and Newman, M.E.J. 2008. Hierarchical structure and the prediction of missing links in networks. *Nature*, **453**(7191), 98–101.

[54] Cormen, T.H., Leiserson, C.E., Rivest, R.L., and Stein, C. 2003. *Introduction to Algorithms*. Cambridge, MA: MIT Press.

[55] Cornea, E.l., Zhu, H., Kim, P., and Ibrahim, J.G. 2016. Regression models on Riemannian symmetric spaces. *Journal of the Royal Statistical Society: Series B (Statistical Methodology)* (in press).

[56] Cox, D.R. 1958. *Planning of Experiments*. New York: Wiley.

[57] Crane, H., and Dempsey, W. 2015. A framework for statistical network modeling. *arXiv preprint arXiv:1509.08185*.

[58] Crane, H., and Dempsey, W. 2016. Edge exchangeable models for network data. *arXiv preprint arXiv:1603.04571*.

[59] Daudin, J.-J., Picard, F., and Robin, S. 2008. A mixture model for random graphs. *Statistics and Computing*, **18**(2), 173–183.

[60] De Nooy, W., Mrvar, A., and Batagelj, V. 2011. *Exploratory Social Network Analysis with Pajek*, Vol. 27. Cambridge: Cambridge University Press.

[61] Dempster, A.P., Laird, N.M., and Rubin, D.B. 1977. Maximum likelihood from incomplete data via the EM algorithm. *Journal of the Royal Statistical Society: Series B (Methodological)*, 39(1), 1–38.

[62] Diaconis, P., and Janson, S. 2008. Graph limits and exchangeable random graphs. *Rendiconti di Matematica, Serie VII*, **28**, 33–61.

[63] Diestel, R. 2005. *Graph Theory*, 3rd edn. Heidelberg: Springer-Verlag.

[64] Dodds, P.S., Muhamad, R., and Watts, D.J. 2003. An experimental study of search in global social networks. *Science*, **301**(5634), 827–829.

[65] DuBois, C., Butts, C.T., and Smyth, P. 2013. Stochastic blockmodeling of relational event dynamics. *Proceedings of the 16th International Conference on Artificial Intelligence and Statistics (AISTATS)*.

[66] Durante, D., Dunson, D.B., and Vogelstein, J.T. 2016. Nonparametric Bayes modeling of populations of networks. *Journal of the American Statistical Association* (in press).

[67] Durrett, R. 2007. *Random Graph Dynamics*. Cambridge: Cambridge University Press.

[68] Eckles, D., Karrer, B., and Ugander, J. 2014. Design and analysis of experiments in networks: Reducing bias from interference. *arXiv preprint arXiv:1404.7530*.

[69] Eldar, Y.C. 2009. Generalized SURE for exponential families: Applications to regularization. *IEEE Transactions on Signal Processing*, **57**(2), 471–481.

[70] Eppstein, D., and Wang, J. 2004. Fast approximation of centrality. *Journal of Graph Algorithms and Applications*, **8**(1), 39–45.

[71] Erdös, P., and Rényi, A. 1959. On random graphs. *Publicationes Mathematicae*, **6**(290), 290–297.

[72] Erdös, P., and Rényi, A. 1960. On the evolution of random graphs. *Publications of the Mathematical Institute of the Hungarian Academy of Sciences*, **5**, 17–61.

[73] Erdös, P., and Rényi, A. 1961. On the strength of connectedness of a random graph. *Acta Mathematica Hungarica*, **12**, 261–267.

[74] Fienberg, S.E. 2012. A brief history of statistical models for network analysis and open challenges. *Journal of Computational and Graphical Statistics*, **21**(4), 825–839.

[75] Forastiere, L., Airoldi, E.M., and Mealli, F. 2016. Identification and estimation of treatment and interference effects in observational studies on networks. *arXiv preprint arXiv:1609.06245*.

[76] Frank, O. 1977. Estimation of graph totals. *Scandinavian Journal of Statistics*, **4**, 81–89.

[77] Frank, O. 1978a. Estimation of the number of connected components in a graph by using a sampled subgraph. *Scandinavian Journal of Statistics*, **5**, 177–188.

[78] Frank, O. 1978b. Sampling and estimation in large social networks. *Social Networks*, **1**(1), 91–101.

[79] Frank, O. 1980. Estimation of the number of vertices of different degrees in a graph. *Journal of Statistical Planning and Inference*, **4**(1), 45–50.

[80] Frank, O. 1981. A survey of statistical methods for graph analysis. *Sociological Methodology*, **12**, 110–155.

[81] Frank, O. 2004. Network sampling and model fitting. In: Carrington, P.J., Scott, J., and Wasserman, S. (eds), *Models and Methods in Social Network Analysis*. New York: Cambridge University Press.

[82] Frank, O., and Harary, F. 1982. Cluster inference by using transitivity indices in empirical graphs. *Journal of the American Statistical Association*, **77**(380), 835–840.

[83] Frank, O., and Strauss, D. 1986. Markov graphs. *Journal of the American Statistical Association*, **81**(395), 832–842.

[84] Frieze, A., and Kannan, R. 1999. Quick approximation to matrices and applications. *Combinatorica*, **19**(2), 175–220.

[85] Fu, W., Song, L., and Xing, E.P. 2009. Dynamic mixed membership block-model for evolving networks. *Proceedings of the 26th International Conference on Machine Learning*, 329–336.

[86] Ganguly, A., and Kolaczyk, E.D. *Estimation of vertex degrees in a sampled network*. arXiv preprint arXiv:1701.07203.

[87] Gao, C., Lu, Y., and Zhou, H.H. 2015. Rate-optimal graphon estimation. *Annals of Statistics*, **43**(6), 2624–2652.

[88] Gile, K.J. 2012. Improved inference for respondent-driven sampling data with application to HIV prevalence estimation. *Journal of the American Statistical Association*, **106**, 135–146.

[89] Gile, K.J., and Handcock, M.S. 2010. Respondent-driven sampling: An assessment of current methodology. *Sociological Methodology*, **40** (1), 285–327.

[90] Gile, K.J., and Handcock, M.S. 2015. Network model-assisted inference from respondent-driven sampling data. *Journal of the Royal Statistical Society: Series A (Statistics in Society)*, **178**(3), 619–639.

[91] Ginestet, C.E., Li, J., Balanchandran, P., Rosenberg, S., and Kolaczyk, E.D. 2017. Hypothesis testing for network data in functional neuroimaging. *Annals of Applied Statistics* (accepted for publication).

[92] Gjoka, M., Kurant, M., Butts, C.T., and Markopoulou, A. 2010. Walking in Facebook: A case study of unbiased sampling of OSNs. *IEEE INFOCOM*, 1–9.

[93] Goel, S., and Salganik, M.J. 2010. Assessing respondent-driven sampling. *Proceedings of the National Academy of Sciences*, **107**(15), 6743–6747.

[94] Gowers, W.T. 1997. Lower bounds of tower type for Szemerédi's uniformity lemma. *Geometric & Functional Analysis*, **7**(2), 322–337.

[95] Granovetter, M. 1976. Network sampling: Some first steps. *American Journal of Sociology*, 81(6), 1287–1303.

[96] Gross, J.L., and Yellen, J. 1999. *Graph Theory and Its Applications*. Boca Raton, FL: Chapman & Hall/CRC.

[97] Guare, J. 1990. *Six Degrees of Separation: A Play*. New York: Vintage.

[98] Halloran, M.E., and Struchiner, C.J. 1995. Causal inference in infectious diseases. *Epidemiology*, **6**(2), 142–151.

[99] Han, J.D.J., Dupuy, D., Bertin, N., Cusick, M.E., and Vidal, M. 2005. Effect of sampling on topology predictions of protein–protein interaction networks. *Nature Biotechnology*, **23**, 839–844.

[100] Handcock, M.S. 2003. *Assessing degeneracy in statistical models of social networks*. Technical Report No. 39 of the Center for Statistics and the Social Sciences, University of Washington.

[101] Handcock, M.S., and Gile, K.J. 2010. Modeling social networks from sampled data. *Annals of Applied Statistics*, **4**(1), 5.

[102] Hart, G.T., Ramani, A.K., and Marcotte, E.M. 2006. How complete are current yeast and human protein–interaction networks? *Genome Biology*, **7**(11), 1.

[103] Heckathorn, D.D. 1997. Respondent-driven sampling: A new approach to the study of hidden populations. *Social Problems*, **44**(2), 174–199.

[104] Heckathorn, D.D. 2002. Respondent-driven sampling II: Deriving valid population estimates from chain-referral samples of hidden populations. *Social Problems*, **49**(1), 11–34.

[105] Ho, Q., Parikh, A.P., and Xing, E.P. 2012. Multiscale community blockmodel for network exploration. *Journal of the American Statistical Association*, **107**(499), 916–934.

[106] Ho, Q., Yin, J., and Xing, E.P. 2016. Latent space inference of Internet-scale networks. *Journal of Machine Learning Research*, **17**(78), 1–41.

[107] Hoff, P.D. 2008. Modeling homophily and stochastic equivalence in symmetric relational data. *Advances in Neural Information Processing Systems (NIPS)*.

[108] Holland, P.W. 1986. Statistics and causal inference. *Journal of the American Statistical Association*, **81**(396), 945–960.

[109] Holland, P.W., Laskey, K.B., and Leinhardt, S. 1983. Stochastic blockmodels: First steps. *Social Networks*, **5**(2), 109–137.

[110] Hoover, D.N. 1979. Relations on probability spaces and arrays of random variables. *Preprint, Institute for Advanced Study, Princeton, NJ*.

[111] Horvitz, D.G., and Thompson, D.J. 1952. A generalization of sampling without replacement from a finite universe. *Journal of the American Statistical Association*, **47**(260), 663–685.

[112] Hübler, C., Kriegel, H.-P., Borgwardt, K., and Ghahramani, Z. 2008. Metropolis algorithms for representative subgraph sampling.

Proceedings of the 8th IEEE International Conference on Data Mining, 283–292.

[113] Hudgens, M.G., and Halloran, M.E. 2012. Toward causal inference with interference. *Journal of the American Statistical Association*, **103**(482), 832–842.

[114] Imbens, G.W., and Rubin, D.B. 2015. *Causal Inference in Statistics, Social, and Biomedical Sciences*. Cambridge: Cambridge University Press.

[115] Jackson, M.O. 2008. *Social and Economic Networks*. Princeton, NJ: Princeton University Press.

[116] Jiang, Q., Zhang, Y., and Sun, M. 2009. Community detection on weighted networks: A variational Bayesian method. *Proceedings of the Asian Conference on Machine Learning*, 176–190.

[117] Jiang, X., and Kolaczyk, E.D. 2012. A latent eigenprobit model with link uncertainty for prediction of protein–protein interactions. *Statistics in Biosciences*, **4**(1), 84–104.

[118] Jiang, X., Gold, D., and Kolaczyk, E.D. 2011. Network-based auto-probit modeling for protein function prediction. *Biometrics*, **67** (3), 958–966.

[119] Johnston, L.G., Hakim, A.J., Dittrich, S., Burnett, J., Kim, E., and White, R.G. 2016. A systematic review of published respondent-driven sampling surveys collecting behavioral and biologic data. *AIDS and Behavior*, **20**(8), 1754–1776.

[120] Jordan, M.I., Ghahramani, Z., Jaakkola, T.S., and Saul, L.K. 1999. An introduction to variational methods for graphical models. *Machine Learning*, **37**(2), 183–233.

[121] Kallenberg, O. 1999. Multivariate sampling and the estimation problem for exchangeable arrays. *Journal of Theoretical Probability*, **12**(3), 859–883.

[122] Kallenberg, O. 2006. *Probabilistic Symmetries and Invariance Principles*. Berlin: Springer Science & Business Media.

[123] Karrer, B., and Newman, M.E.J. 2011. Stochastic blockmodels and community structure in networks. *Physical Review E*, **83**(1), 016107.

[124] Keele, L. 2015. The statistics of causal inference: A view from political methodology. *Political Analysis*, **23**(3), 313–335.

[125] Killworth, P.D., McCarty, C., Bernard, H.R., Shelley, G.A., and Johnsen, E.C. 1998. Estimation of seroprevalence, rape, and homelessness in the United States using a social network approach. *Evaluation Review*, **22**(2), 289–308.

[126] Kohavi, R., and Longbotham, R. 2015. Online controlled experiments and A/B tests. In: Sammut, C., and Webb, G. (eds), *Encyclopedia of Machine Learning and Data Mining*. Berlin: Springer-Verlag.

[127] Kolaczyk, E.D. 2009. *Statistical Analysis of Network Data: Methods and Models*. Berlin: Springer-Verlag.

[128] Kolaczyk, E.D., and Csárdi, G. 2014. *Statistical Analysis of Network Data with R*. Berlin: Springer-Verlag.

[129] Kolar, M., Liu, H., and Xing, E.P. 2014. Graph estimation from multi-attribute data. *Journal of Machine Learning Research*, **15**(1), 1713–1750.

[130] Koutsourelakis, P.-S., and Eliassi-Rad, T. 2008. Finding mixed-memberships in social networks. *AAAI Spring Symposium: Social Information Processing*, 48–53.

[131] Krivitsky, P.N., and Kolaczyk, E.D. 2015. On the question of effective sample size in network modeling: An asymptotic inquiry. *Statistical Science*, **30**(2), 184.

[132] Lakhina, A., Byers, J.W., Crovella, M., and Xie, P. 2003. Sampling biases in IP topology measurements. *IEEE INFOCOM*, 332–341.

[133] Lee, S.H., Kim, P.J., and Jeong, H. 2006. Statistical properties of sampled networks. *Physical Review E*, **73**(1), 16102.

[134] Lei, J. 2016. A goodness-of-fit test for stochastic block models. *Annals of Statistics*, **44**(1), 401–424.

[135] Leskovec, J., and Faloutsos, C. 2006. Sampling from large graphs. *Proceedings of the 12th ACM International Conference on Knowledge Discovery and Data Mining (SIGKDD)*, 631–636.

[136] Leung, M. 2016. Treatment and spillover effects under network interference. Available at: ssrn 2757313.

[137] Li, X., and Rohe, K. 2015. Central limit theorems for network-driven sampling. *arXiv preprint arXiv:1509.04704*.

[138] Lovász, L. 2012. *Large Networks and Graph Limits*, Vol. 60. American Mathematical Society.

[139] Lovász, L., and Szegedy, B. 2006. Limits of dense graph sequences. *Journal of Combinatorial Theory, Series B*, **96**(6), 933–957.

[140] Lu, X., Bengtsson, L., Britton, T., Camitz, M., Kim, B.J., Thorson, A., and Liljeros, F. 2012. The sensitivity of respondent-driven sampling. *Journal of the Royal Statistical Society: Series A (Statistics in Society)*, **175**(1), 191–216.

[141] Lu, X., Malmros, J., Liljeros, F., and Britton, T. 2013. Respondent-driven sampling on directed networks. *Electronic Journal of Statistics*, 7, 292–322.

[142] Lunagomez, S., and Airoldi, E.M. 2014. Valid inference from non-ignorable network sampling designs. *arXiv preprint arXiv:1401.4718*.

[143] Lusher, D., Koskinen, J., and Robins, G.L. 2012. *Exponential Random Graph Models for Social Networks: Theory, Methods, and Applications*. Cambridge: Cambridge University Press.

[144] Lyons, R. 2011. The spread of evidence-poor medicine via flawed social-network analysis. *Statistics, Politics, and Policy*, **2**(1).

[145] Manski, C.F. 1993. Identification of endogenous social effects: The reflection problem. *Review of Economic Studies*, **60**(3), 531–542.

[146] Manski, C.F. 2013. Identification of treatment response with social interactions. *Econometrics Journal*, **16**(1), S1–S23.

[147] Mariadassou, M., Robin, S., and Vacher, C. 2010. Uncovering latent structure in valued graphs: A variational approach. *Annals of Applied Statistics*, **4**(2), 715–742.

[148] Matias, C., and Miele, V. 2017. Statistical clustering of temporal networks through a dynamic stochastic block model. *Journal of the Royal Statistical Society: Series B (Statistical Methodology)* (in press).

[149] Matias, C., and Robin, S. 2014. Modeling heterogeneity in random graphs through latent space models: A selective review. *ESAIM: Proceedings and Surveys*, **47**, 55–74.

[150] McCormick, T.H., Salganik, M.J., and Zheng, T. 2010. How many people do you know? Efficiently estimating personal network size. *Journal of the American Statistical Association*, **105**(489), 59–70.

[151] Meinshausen, N., and Bühlmann, P. 2006. High-dimensional graphs and variable selection with the Lasso. *Annals of Statistics*, **34**(3), 1436–1462.

[152] Michael, J.H. 1997. Labor dispute reconciliation in a forest products manufacturing facility. *Forest Products Journal*, **47**(11/12), 41.

[153] Milgram, S. 1967. The small world problem. *Psychology Today*, **2**(1), 60–67.

[154] Newman, M.E.J. 2010. *Networks: An Introduction*. Oxford: Oxford University Press.

[155] Nowicki, K., and Snijders, T.A.B. 2001. Estimation and prediction for stochastic blockstructures. *Journal of the American Statistical Association*, **96**(455), 1077–1087.

[156] Ogburn, E.L., and VanderWeele, T.J. 2014. Causal diagrams for interference. *Statistical Science*, **29**(4), 559–578.

[157] Olhede, S.C., and Wolfe, P.J. 2014. Network histograms and universality of blockmodel approximation. *Proceedings of the National Academy of Sciences*, **111**(41), 14722–14727.

[158] Orbanz, P., and Roy, D.M. 2015. Bayesian models of graphs, arrays and other exchangeable random structures. *IEEE Transactions on Pattern Analysis and Machine Intelligence*, **37**(2), 437–461.

[159] Paluck, E.L. 2011. Peer pressure against prejudice: A high school field experiment examining social network change. *Journal of Experimental Social Psychology*, **47**(2), 350–358.

[160] Park, J., and Newman, M.E.J. 2004. Solution of the two-star model of a network. *Physical Review E*, **70**(6), 066146.

[161] Priebe, C.E., Sussman, D.L., Tang, M., and Vogelstein, J.T. 2015. Statistical inference on errorfully observed graphs. *Journal of Computational and Graphical Statistics*, **24**(4), 930–953.

[162] Ramani, S., Blu, T., and Unser, M. 2008. Monte-Carlo SURE: A black-box optimization of regularization parameters for general denoising algorithms. *IEEE Transactions on Image Processing*, **17**(9), 1540–1554.

[163] Robins, G.L., and Morris, M. 2007. Advances in exponential random graph (p*) models. *Social Networks*, **29**(2), 169–172.

[164] Robins, G.L., Pattison, P.E., Kalish, Y., and Lusher, D. 2007a. An introduction to exponential random graph (p*) models for social networks. *Social Networks*, **29**(2), 173–191.

[165] Robins, G.L., Snijders, T., Wang, P., Handcock, M., and Pattison, P.E. 2007b. Recent developments in exponential random graph (p*) models for social networks. *Social Networks*, **29**(2), 192–215.

[166] Rohe, K., Chatterjee, S., and Yu, B. 2011. Spectral clustering and the high-dimensional stochastic blockmodel. *Annals of Statistics*, **39**(4), 1878–1915.

[167] Ron, D. 2001. Property testing. *Journal of Combinatorial Optimization*, **9**(2), 597–643.

[168] Rosenbaum, P.R. 1999. Reduced sensitivity to hidden bias at upper quantiles in observational studies with dilated treatment effects. *Biometrics*, **5**(2), 560–564.

[169] Rosenbaum, P.R. 2007. Interference between units in randomized experiments. *Journal of the American Statistical Association*, **102**(477), 191–200.

[170] Ross, R. 1916. An application of the theory of probabilities to the study of a priori pathometry. Part I. *Proceedings of the Royal Society of London. Series A, Containing Papers of a Mathematical and Physical Character*, **92**(638), 204–230.

[171] Rubin, D.B. 1974. Estimating causal effects of treatments in randomized and nonrandomized studies. *Journal of Educational Psychology*, **66**(5), 688.

[172] Rubin, D.B. 1990. [On the Application of Probability Theory to Agricultural Experiments. Essay on Principles. Section 9.] Comment: Neyman (1923) and Causal Inference in Experiments and Observational Studies. *Statistical Science*, **5**(4), 472–480.

[173] Salganik, M.J., and Heckathorn, D.D. 2004. Sampling and estimation in hidden populations using respondent-driven sampling. *Sociological Methodology*, **34**(1), 193–240.

[174] Schweinberger, M. 2011. Instability, sensitivity, and degeneracy of discrete exponential families. *Journal of the American Statistical Association*, **106**(496), 1361–1370.

[175] Schweinberger, M., and Handcock, M.S. 2015. Local dependence in random graph models: Characterization, properties and statistical inference. *Journal of the Royal Statistical Society: Series B (Statistical Methodology)*, **77**(3), 647–676.

[176] Shalizi, C.R., and Rinaldo, A. 2013. Consistency under sampling of exponential random graph models. *Annals of Statistics*, **41**(2), 508–535.

[177] Shalizi, C.R., and Thomas, A.C. 2011. Homophily and contagion are generically confounded in observational social network studies. *Sociological Methods & Research*, **40**(2), 211–239.

[178] Shuai, H.-H., Yang, D.-N., Shen, C.-Y., Philip, S.Y., and Chen, M.-S. 2015. QMSampler: Joint sampling of multiple networks with quality guarantee. *arXiv preprint arXiv:1502.07439*.

[179] Snijders, T.A.B., and Nowicki, K. 1997. Estimation and prediction for stochastic blockmodels for graphs with latent block structure. *Journal of Classification*, **14**(1), 75–100.

[180] Sobel, M.E. 2006. What do randomized studies of housing mobility demonstrate? Causal inference in the face of interference. *Journal of the American Statistical Association*, **101**(476), 1398–1407.

[181] Söderberg, B. 2002. General formalism for inhomogeneous random graphs. *Physical Review E*, **66**(6), 066121.

[182] Söderberg, B. 2003a. Properties of random graphs with hidden color. *Physical Review E*, **68**(2), 026107.

[183] Söderberg, B. 2003b. Random graphs with hidden color. *Physical Review E*, **68**(1), 015102.

[184] Solomon, P., Cavanaugh, M.M., and Draine, J. 2009. *Randomized Controlled Trials: Design and Implementation for Community-Based Psychosocial Interventions*. Oxford: Oxford University Press.

[185] Stanley, N., Shai, S., Taylor, D., and Mucha, P. 2015. Clustering network layers with the strata multilayer stochastic block model. *IEEE Transactions on Network Science and Engineering*, **3**(2), 95–105.

[186] Stanley, N., Shai, S., Taylor, D., and Mucha, P. 2016. Clustering network layers with the strata multilayer stochastic block model. *IEEE Transactions on Network Science and Engineering*, **3**(2), 95–105.

[187] Stumpf, M.P.H., and Wiuf, C. 2005. Sampling properties of random graphs: The degree distribution. *Physical Review E*, **72**(3), 36118.

[188] Stumpf, M.P.H., Wiuf, C., and May, R.M. 2005. Subnets of scale-free networks are not scale-free: Sampling properties of networks. *Proceedings of the National Academy of Sciences*, **102**(12), 4221–4224.

[189] Sussman, D.L., Tang, M., Fishkind, D.E., and Priebe, C.E. 2012. A consistent adjacency spectral embedding for stochastic blockmodel graphs. *Journal of the American Statistical Association*, **107**(499), 1119–1128.

[190] Sweet, T.M. 2015. Incorporating covariates into stochastic blockmodels. *Journal of Educational and Behavioral Statistics*, **40**(6), 635–664.

[191] Tallberg, C. 2004. A Bayesian approach to modeling stochastic blockstructures with covariates. *Journal of Mathematical Sociology*, **29**(1), 1–23.

[192] Tang, M., Sussman, D.L., and Priebe, C.E. 2013. Universally consistent vertex classification for latent positions graphs. *Annals of Statistics*, **41** (3), 1406–1430.

[193] Tang, R., Ketcha, M., Vogelstein, J.T., Priebe, C.E., and Sussman, D.L. 2016. Law of large graphs. *arXiv preprint arXiv:1609.01672*.

[194] Tchetgen, E.J.T., and VanderWeele, T.J. 2012. On causal inference in the presence of interference. *Statistical Methods in Medical Research*, **21**(1), 55–75.

[195] Thompson, S.K. 2002. *Sampling*, 2nd edn. New York: Wiley.

[196] Thompson, S.K., and Frank, O. 2000. Model-based estimation with link-tracing sampling designs. *Survey Methodology*, **26**(1), 87–98.

[197] Toulis, P., and Kao, E.K. 2013. Estimation of causal peer influence effects. *Proceedings of the 30th International Conference on Machine Learning*, 1489–1497.

[198] Ugander, J., Karrer, B., Backstrom, L., and Kleinberg, J. 2013. Graph cluster randomization: Network exposure to multiple universes. *Proceedings of the 19th ACM International Conference on Knowledge Discovery and Data Mining (SIGKDD)*, 329–337.

[199] Veitch, V., and Roy, D.M. 2015. The class of random graphs arising from exchangeable random measures. *arXiv preprint arXiv:1512.03099*.

[200] Volz, E., and Heckathorn, D.D. 2008. Probability based estimation theory for respondent driven sampling. *Journal of Official Statistics*, **24**(1), 79.

[201] Vu, D.Q., Hunter, D.R., and Schweinberger, M. 2013. Model-based clustering of large networks. *Annals of Applied Statistics*, **7**(2), 1010.

[202] Wang, H., and Marron, J.S. 2007. Object oriented data analysis: Sets of trees. *Annals of Statistics*, **35**(5), 1849–1873.

[203] Wang, Y.J., and Wong, G.Y. 1987. Stochastic blockmodels for directed graphs. *Journal of the American Statistical Association*, **82**(397), 8–19.

[204] Wasserman, S., and Pattison, P.E. 1996. Logit models and logistic regressions for social networks: I. An introduction to Markov graphs and p^*. *Psychometrika*, **61**(3), 401–425.

[205] Wolfe, P.J., and Olhede, S.C. 2013. Nonparametric graphon estimation. *arXiv preprint arXiv:1309.5936*.

[206] Xu, K.S., and Hero, A.O. 2014. Dynamic stochastic blockmodels for time-evolving social networks. *IEEE Journal of Selected Topics in Signal Processing*, **8**(4), 552–562.

[207] Yang, J., and Leskovec, J. 2012. Defining and evaluating network communities based on ground-truth. *Proceedings of the ACM SIGKDD Workshop on Mining Data Semantics*, 3.

[208] Yang, J., Han, C., and Airoldi, E.M. 2014. Nonparametric estimation and testing of exchangeable graph models. *AISTATS*, 1060–1067.

[209] Yang, T., Chi, Y., Zhu, S., Gong, Y., and Jin, R. 2011. Detecting communities and their evolutions in dynamic social networks – a Bayesian approach. *Machine Learning*, **82**(2), 157–189.

[210] Young, S.J., and Scheinerman, E.R. 2007. Random dot product graph models for social networks. *International Workshop on Algorithms and Models for the Web-Graph*, 138–149.

[211] Zhang, C.-H. 2005. Estimation of sums of random variables: Examples and information bounds. *Annals of Statistics*, **33**(5), 2022–2041.

[212] Zhang, Y., Kolaczyk, E.D., and Spencer, B.D. 2015a. Estimating network degree distributions under sampling: An inverse problem, with applications to monitoring social media networks. *Annals of Applied Statistics*, **9**(1), 166–199.

[213] Zhang, Y., Levina, E., and Zhu, J. 2015b. Estimating network edge probabilities by neighborhood smoothing. *arXiv preprint arXiv:1509.08588*.

[214] Zhao, Y., Levina, E., and Zhu, J. 2012. Consistency of community detection in networks under degree-corrected stochastic block models. *Annals of Statistics*, **40**(4), 2266–2292.

Author Index

Subject Index

Printed in the United States
By Bookmasters